Textbooks in Telecommunication Engineering

Era of the Internet and Network Science/Engineering

Series Editor

Tarek S. El-Bawab, Boston, MA, USA

Telecommunication and networks have evolved to embrace all aspects of our everyday life, including smart cities and infrastructures, healthcare, banking and businesses, manufacturing, space and aviation, meteorology and climate change, oceans and marine life, Internet of Things, defense, homeland security, education, research, social media, entertainment, and many others. Network applications and services continue to expand, virtually without limits. Therefore, specialized telecommunication and network engineering programs are recognized as a necessity to accelerate the pace of advancement in this field, and to prepare a new generation of engineers for imminent needs in our modern life. These programs need curricula, courses, labs, and textbooks of their own.

The IEEE Communications Society's Telecommunication Engineering Education (TEE) movement, led by Tarek S. El-Bawab- the editor of this Series, resulted in recognition of this field of engineering by the Accreditation Board for Engineering and Technology (ABET) on November 1, 2014. This Springer Series was launched to capitalizes on this milestone. The Series goal is to produce high-quality textbooks to fulfill the education needs of telecommunication and network engineering, and to support the development of specialized undergraduate and graduate curricula in this regard. The Series also supports research in this field and helps prepare its scholars for global challenges that lay ahead. The Series have published innovative textbooks in areas of network science and engineering where textbooks have been rare. It is producing high-quality volumes featuring innovative presentation media, interactive content, and online resources for students and professors.

Book proposals are solicited in all topics of telecommunication and network engineering including, but not limited to: network architecture and protocols; traffic engineering; network design, dimensioning, modeling, measurements, and analytics; network management and softwarization; cybersecurity; synchronization and control; applications of artificial intelligence in telecommunications and networks; applications of data science in telecommunications and networks; network availability, reliability, protection, recovery and restoration; wireless communication systems; cellular technologies and networks (through 5G, 6G, and beyond); satellite and space communications and networks; optical communications and networks; heterogeneous networks; broadband access and free-space optical communications; MSO/cable networks; storage networks; optical interconnects; and data centers; social networks; transmission media and systems; switching and routing (from legacy to today's paradigms); network applications and services; telecom economics and business; telecom regulation and policies; standards and standardization; and laboratories.

Proposals of interest shall be for textbooks that can be used to develop university courses, either in full or in part. They should include recent advances in the field while capturing whatever fundamentals that are necessary for students to understand the bases of the topic and appreciate its evolution trends. Books in this series will provide high-quality illustrations, examples, end-of-chapters' problems/exercises and case studies.

This series is indexed in Scopus. For further information and to submit proposals, please contact the Series Editor, Dr. Tarek S. El-Bawab, telbawab@ieee.org; or Mary James, Executive Editor at Springer, mary.james@springer.com.

Masatoshi Suzuki · Itsuro Morita

Long-Haul Optical Fiber Communication Systems

Springer

Masatoshi Suzuki
Chitose Institute of Science and Technology
Chitose, Japan

Itsuro Morita
Waseda University
Tokyo, Japan

This book also provides accompanying materials for lecturers. You can request them by going to sn.pub/lecturermaterial and filling out the form.

ISSN 2524-4345 ISSN 2524-4353 (electronic)
Textbooks in Telecommunication Engineering
ISBN 978-3-032-19471-8 ISBN 978-3-032-19472-5 (eBook)
https://doi.org/10.1007/978-3-032-19472-5

This Springer imprint is published by the registered company Springer Nature Switzerland AG
The registered company address is: Gewerbestrasse 11, 6330 Cham, Switzerland

Note from the Series Editor

This Series has its roots in the Telecommunication Engineering Education (TEE) movement in the USA, which I had the honor of leading from 2008 to 2015. The movement was a result of remarkable impacts telecom networks had already started to have on society, and many more that were looming on the horizon at that time. Major changes in the theoretical foundations of this field of engineering, its design strategies, and its practical applications were unfolding at the same time. The industries we had known as the Telecom, Cable, and Data-Communication industries through the twentieth century were going through technical innovations and were reshuffling, repositioning themselves in the market, and converging, in preparation for the global economy of the twenty-first century.

The TEE movement was prompted by a personal experience I went through as I moved from industry to academia. Telecom engineering, or Network Engineering, since the network is centric to the evolution telecom has been through for three to four decades so far, became a distinct multidisciplinary field of study. This field blends ingredients from electrical engineering, computer engineering, computer science, and other disciplines, but may not be taught merely as a branch of any one of them, per se. It became evident that society would be better served by adding specialized undergraduate/graduate courses and programs in telecom/network engineering to our higher-education system, especially in the USA where this was not the norm.

The story of the TEE movement has been told and also discussed within broader contexts in the literature (e.g., IEEE Communications Magazine, Vol. 53, No. 11, pp. 35–9, November 2015). The movement had accomplished its main goal of adding telecom/network engineering to ABET's portfolio of recognized engineering disciplines. ABET is the Accreditation Board of Engineering and Technology, Inc., which accredits engineering programs in the USA and more than 40 other countries. A lot of work had to be done to capitalize on that milestone, however. The technical literature of this discipline was rich in scholarly research papers, but not as rich in textbooks designed to support the development of new university courses and curricula. These textbooks are needed to facilitate rigorous academic training for a new generation of network engineers. The idea of this Series came up subsequently to help the telecom/networking community overcome this shortcoming. Since its inception, Springer's

TTE (Textbooks in Telecommunication Engineering) has been doing this job. The proximity of the acronyms herein—between TEE and TTE—is coincidental, but worth noting!

I am proud of the inroads TTE has made in publishing quality textbooks and in fulfilling the goals and objectives the Series has had since its beginning. By the end of 2024, we approach the milestone of having 35 books, including 20 published and about 15 in the making (whether in the writing, reviewing, or production phases). More than 50 authors have contributed to this Series so far and they come from about 20 different countries and cultures all over the world.

The topics covered by this Series so far scan the entire landscape of telecom and network engineering, including wireless, optical, satellite; access and core; devices, systems, and networks; space-aerial-terrestrial integration; queuing theory and traffic engineering; network management; Internet and Internet of Things (IoT); signal processing; applied electromagnetics; and economy/business aspects of telecom. The Series includes a cellulargenerations primer and has a Wi-Fi primer project under progress. It covers the most cutting-edge topics with textbooks that are carefully designed to strike the right balance between the need to create new courses thereto today and the fact that progress is imminent in each of these topics, and in the short term. These include books that discuss the roles of edge and cloud computing; of blockchains, and of artificial intelligence and machine learning in network engineering. Special topics such as under-water communications, railways telecom, networked navigation systems, semantic communications, video analytics, and broadband to combat the digital divide are also covered by TTE. I have tried my best to make this Series a top de facto global home for quality telecom/network engineering textbooks, and am hoping that many in our community can start to consider it as such now. The fruits of the hard work put by many people into fulfilling this goal are now handy to pick by our community.

Tarek El-Bawab, PhD
December 2024

Preface

Optical communications technologies have been progressing and increased the transmission capacity by more than six orders of magnitude within four decades. The evolution in optical communications technologies has significantly contributed to the worldwide spread of the Internet with rich content and resulted in advancement of today's information society. The coherent light from semiconductor lasers and the low-loss and broadband characteristics of optical fibers were attracting attention from the 1970s, leading to pioneering research on long-distance transmission systems. Among these advancements, the development of optical submarine cable systems and networks, which greatly facilitated the globalization of information communication, represented the pinnacle of technology in this field and gained worldwide attention, achieving remarkable progress.

This book summarizes long-distance, large-capacity optical fiber communication systems based on the experiences of the authors, who have been engaged in research and development in the field of long-distance large-capacity optical communication systems including submarine cable systems for many years. The key driver of technological innovation in this area was not just the optical technology itself but also the emergence of new demands that continually pushed for greater capacity accompanying the rapid growth of the Internet and AI technologies. Therefore, Chap. 1 introduces the background of these developments through an overview of the evolution of long-distance large-capacity optical fiber communication. Chapters 2 and 3 provide concise explanation of the fundamental technologies of optical fiber communication. From Chap. 4 onward, the book covers the transition from optical regeneration repeater systems to optical amplifier repeater systems, the introduction of wavelength division multiplexing, optical fiber dispersion management, and digital coherent technology, all of which enabled truly large-capacity transmission. Finally, the book details Space-Division Multiplexing technology as a next-generation innovation. We believe that this book offers students and researchers in the area of optical fiber communication systems an essential guide to optical transmission technologies.

Finally, we would like to express our gratitude to Dr. Shigeyuki Akiba for sharing his expertise in submarine cable systems and the collaborators in KDDI Research for their continuous team efforts and enthusiasm in an advancing exiting field.

Chitose, Japan Masatoshi Suzuki
Tokyo, Japan Itsuro Morita

Competing Interests The authors have no competing interests to declare that are relevant to the content of this manuscript.

Contents

Chapter 1
Evolution of Communication Technologies

Abstract In this chapter, an overview of the evolution of long-distance communication technologies, specifically from the perspective of submarine cable systems, is provided based on author's extensive experience in research, development, and business in this field.

Keywords Optical communication systems · Optical submarine cable systems · Trans-Pacific cable systems · Power feeding equipment · Submarine cable · Three-divided steel segment

1.1 100 Years of International Communication

The technology of laying cables for communication on the seabed is indeed a specialized area. Due to the significantly different environment from land, designing, manufacturing, and laying cables and repeaters under sea require advanced technologies. Additionally, the difficulty of repair requires exceptionally high reliability. Despite these challenges, the first submarine cable was laid across the Dover Strait between England and France as early as 1851, marking over 150 years of history. Figure 1.1 focuses on Trans-Pacific cables over a 100-year period starting from around 1900, with references to the Dover Strait cable and the first Trans-Atlantic cable before 1900 also included for context. Following the telegraph cables using Morse code, telephone cables were laid in 1891. In the Pacific, a telegraph cable was constructed between Tokyo and Guam in 1906, but the primary method of intercontinental communication later shifted to satellite communication, which used a geostationary satellite as a repeater. In 1964, the first Trans-Pacific cable system (Trans-Pacific Cable-1, TPC-1) using coaxial cables and submarine repeaters with vacuum tube amplifiers was completed, allowing about 100 people to simultaneously talk via voice communication. This was also the year of the Tokyo Olympics, but critical video transmissions still had to rely on satellite communication. After the introduction of the third Trans-Pacific cable system (TPC-3) utilizing optical fibers in 1989, the video services were also provided by submarine cable systems.

M. Suzuki and I. Morita, *Long-Haul Optical Fiber Communication Systems*, Textbooks in Telecommunication Engineering, https://doi.org/10.1007/978-3-032-19472-5_1

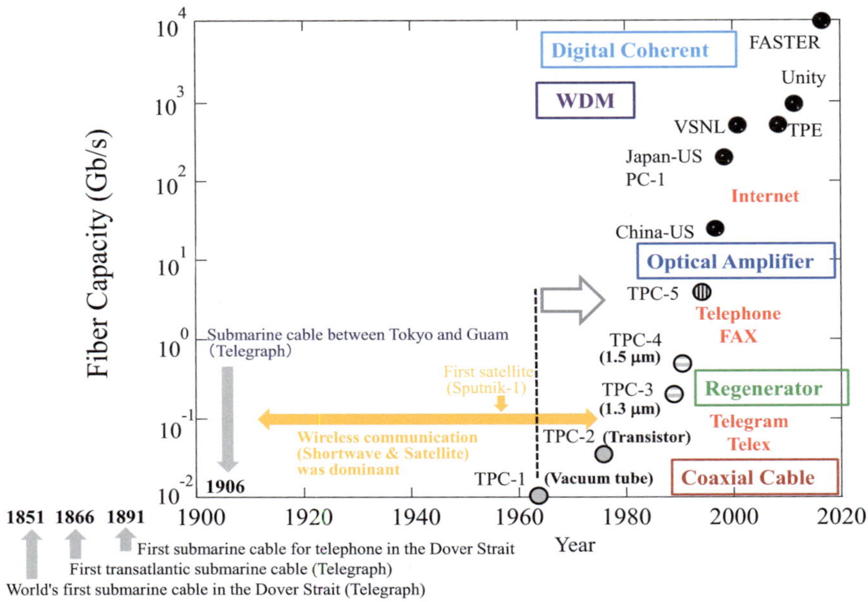

Fig. 1.1 Evolution of long-distance transmission technologies in Trans-Pacific communication

After the introduction of optical fiber, the increase in transmission capacity has been truly remarkable. It grew 1000 times in 10 years, effectively doubling each year on average. Another significant aspect related to submarine cable systems was the changing demands that continuously drove technological innovation, influenced by the evolving social background surrounding communications. The first optical submarine cable system using 1.3 μm lasers, TPC-3, was constructed in 1989. With a fiber capacity of 280 Mb/s, it offered significantly broader bandwidth compared to previous coaxial cable systems and it was believed that no new cable systems would be needed until the end of the twentieth century. However, the demand for telephone services between Japan and the United States grew rapidly, and capacity was strained soon after the cable's completion, leading to the necessity of TPC-4. By utilizing a single-mode laser at 1.5 μm, the fiber capacity doubled to 560 Mb/s, and the repeater spacing extended dramatically from 50 to 138 km, enabling the first direct connection between Japan and the U.S. Despite expectations that this capacity would be sufficient for the foreseeable future, traffic demands surged again shortly after its construction in 1992. This increase was driven by the growing need for not only international telephone services but also international fax communications.

Around 1990, a completely different method from optical regeneration repeater, optical amplification, was put into practical use. This method utilizes Erbium-Doped Fiber pumped by strong light at 1.48 μm or 0.98 μm for amplification in the 1.5 μm band. By repeating such amplification every few tens of kilometers, it became possible to transmit data across the 9000 km of the Pacific Ocean without needing signal regeneration, achieving a large fiber capacity of 5 Gb/s in TPC-5, which was eight times the bit rate of TPC-4. The TPC-5 cable network, which employed optical amplification repeaters, was constructed as a ring network with cables on the northern and southern routes, each backing up the other. It was believed that no further submarine cable systems would be needed indefinitely. This might have been true if the demand was limited to international telephone and fax services.

However, this expectation was once again overturned by the rise of the internet. Internet traffic grew explosively, creating a situation where even numerous new submarine cable systems were insufficient to meet the demand. In response to this, wavelength division multiplexing technology and digital coherent detection technology were brought into practical use, aligning supply with demand and resulting in a dramatic increase in fiber capacity. The fiber capacity from TPC-1 to FASTER cable (with 10 Tb/s) has seen a 6–7 order of magnitude increase.

1.2 Terrestrial Communication Systems and Submarine Cable Systems

1.2.1 Transmission Distances

When we refer to long-distance transmission, we are talking about situations where the distance between the transmitting and receiving terminals is significantly long. In terrestrial cases, this typically refers to transmission systems between major cities. In Japan, for instance, this would include routes like Tokyo to Osaka, covering distances of several hundred kilometers, and at most, around 1000 km. On the other hand, in the cases of submarine cable systems, as illustrated in Fig. 1.2, the distance between the landing stations often spans entire oceans. For example, the distance between Japan and the United States is approximately 9000 km, while the routes from China or the Philippines to the United States can reach up to 12,000 km. Obviously, the longer the distance, the greater the technical challenges, and the more advanced the required technologies become. This requires the integration of the most cutting-edge technologies, bringing together the "best of the best" in the field.

Fig. 1.2 Schematic diagram of an optical submarine cable system

1.2.2 Transmission Costs and Terminal Costs

Advancements in wavelength division multiplexing (WDM) technologies have significantly increased the number of wavelengths that can be transmitted through a single optical fiber. To support WDM transmission, both the optical fiber and repeaters need to meet higher specifications, which can raise costs, though not in direct proportion to the number of wavelengths. In many cases, systems undergo a "upgrade" where the number of wavelengths transmitted is increased beyond the originally design and/or more advanced modulation format is applied to increase the channel bit rate. This is often done without changing the main components of the transmission system, such as the optical fiber and repeaters, meaning the transmission system's costs do not increase correspondingly. On the other hand, when the number of wavelengths increases, the number of transmitters and receivers at the terminal stations must also increase proportionally, leading to higher terminal station costs. Terminal stations have various equipment in addition to transmitters and receivers, so while the costs do not increase linearly with the number of wavelengths, they do grow. Moreover, this increase in equipment may require more floor space, and in some cases, the construction of new facilities may be necessary. In terrestrial systems, the decision to increase the number of wavelengths or to add more optical fibers and repeaters is often a subject of debate, especially when building new systems. Factors such as transmission distance and future capacity needs must be carefully considered. However, in submarine cable systems, where the number of fibers that can be accommodated in a cable is limited, there is a much stronger incentive to increase the number of wavelengths.

1.2.3 Power Supply

In terrestrial systems, power supply is relatively straightforward; equipment that requires electricity is typically installed where power is readily available. However, in submarine cable systems, the power needed by repeaters is supplied through the cable itself. Figure 1.3 shows a photograph and cross-sectional structure of an optical submarine cable. The electric current required by a typical repeater is usually around 1 A. To supply this power, a tubular copper conductor is used, which has a resistance of about 0.7 ~ 1.0 Ω/km. For a system spanning the 9000 km distance across the Pacific Ocean, the voltage drop due to ohmic loss would be around 6300 V or more. This voltage drop must be added to the operating voltage required by the repeaters. For example, if the system length is 9000 km and cable resistance is 0.7 Ω/km and the repeater spacing is 50 km, 180 repeaters are needed. If each repeater requires 20 V, then 180 repeaters would need a total of 3600 V. In this case, the power supply from the power feeding equipment (PFE) at the terminal stations would need to provide a voltage of 9900 V. The maximum voltage that recent PFE can supply is around 18,000 V.

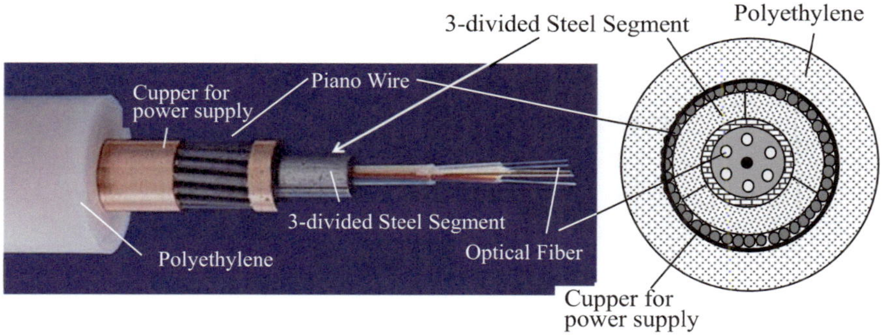

Fig. 1.3 External view photograph and internal cross-sectional structure of an optical submarine cable

To reduce ohmic losses in the cable, one might consider increasing the thickness of the copper used for power supply. However, this would lead to an increase in both the weight and diameter of the cable, which would, in turn, raise the costs of the cable itself and its installation. This is generally undesirable. In Fig. 1.3, the polyethylene layer serves to insulate the power-supplying copper from seawater.

1.2.4 Laying Environment

Land cables and repeaters are typically installed in protected conduits or ducts. In contrast, undersea cables are laid directly on the ocean floor. When crossing the Pacific Ocean, for instance, they traverse the deepest parts of the Japan Trench, which reaches depths of around 8000 m, subjecting the cables to water pressures as high as 800 atmospheric pressures. While repeaters are positioned to avoid the deepest areas, the cables themselves must withstand these extreme depths, enduring up to 800 atmospheric pressures. The three-divided steel segment shown in Fig. 1.3 is designed to withstand such high pressures. Additionally, the installation of undersea cables is carried out using cable ships, during which the cables are subjected to significant tension, measured in tons. To endure this tension, the cables are reinforced with piano wire wound in a helical pattern.

1.2.5 Reliability

Communication systems operate continuously, 24 h a day, for extended periods. Therefore, the equipment and components used must have high reliability, and while the failure rate is extremely low, it is not zero. In the event of a failure, repairing or replacing equipment in terrestrial systems is relatively easy, whereas in submarine cable systems, it involves significant difficulties. Consequently, a stringent reliability design is implemented, such that the system is expected to require no more than three repairs over its operational lifespan, for example, 25 years.

Problems

1.1 How many people can communicate simultaneously by telephone with the fiber capacity of TPC-4 assuming the required bit rate for a telephone service is 64 kb/s?

1.2 How many people can download video content simultaneously with the fiber capacity of TPC-4 assuming the required bit rate for downloading the video content is 2 Mb/s?

1.3 Calculate the required voltage for supplying power from the terminal stations to a Trans-Pacific submarine cable system with the distance of 9000 km in the following condition.

- Resistance of tubular copper conductor: 1 Ω/km.
- Electric current: 1 A.
- Repeater spacing: 50 km.
- Required voltage in each repeater: 18 V.

1.4 Calculate the required voltage for supplying power from the terminal stations to a Trans-Atlantic submarine cable system with the distance of 6000 km in the same condition as 1.1.

Chapter 2
Key Technologies in Optical Fiber Communication Systems

Abstract This chapter outlines the key technologies required for long-distance optical fiber communication systems. In the optical transmitter, the semiconductor laser driven by the electrical signal converts it to an optical signal. Optical modulators may also be used for generating optical signals. The transmitted optical signal is attenuated by losses in the optical fiber, and the optical repeater restores the optical signal level to its original state before sending it to the next optical fiber section. Optical repeaters include regeneration repeaters, which reproduce optical signals through optical-electrical-optical conversion, and optical amplifier repeaters, which amplify optical signals directly in the optical domain. On the receiver side, the transmitted optical signal is converted into an electrical signal by the photodetector, and the TDM demultiplexing circuit demultiplexes to the low-speed electrical signals, transmitting the information to the destination.

Keywords Single-mode fiber · Semiconductor laser · Optical modulator · Photo detector · Optical amplifier · Optical repeater · Forward error correction

Figure 2.1 shows the basic configuration of an optical fiber communication system that uses the Intensity Modulation-Direct Detection (IM-DD) method. On the transmission side, low-speed electrical signals, which are the source information, are multiplexed into high-speed digital electrical signals by the Time Division Multiplexing (TDM) circuit. In long-distance optical fiber communication systems, the signals are usually multiplexed in multiples of four (e.g., from 150 to 600 Mb/s, 2.4 Gb/s, etc.) according to the hierarchical levels of the Optical Transport Network (OTN), the standard signal format for optical networks. In the optical transmitter, the semiconductor laser driven by the electrical signal converts it to an optical signal. Optical modulators may also be used for generating optical signals. The transmitted optical signal is attenuated by losses in the optical fiber, and the optical repeater restores the optical signal level to its original state before sending it to the next optical fiber section. Optical repeaters include regenerative repeaters, which reproduce optical signals through optical–electrical–optical conversion, and optical amplifier repeaters, which amplify optical signals directly in the optical domain. On the

M. Suzuki and I. Morita, *Long-Haul Optical Fiber Communication Systems*, Textbooks in Telecommunication Engineering, https://doi.org/10.1007/978-3-032-19472-5_2

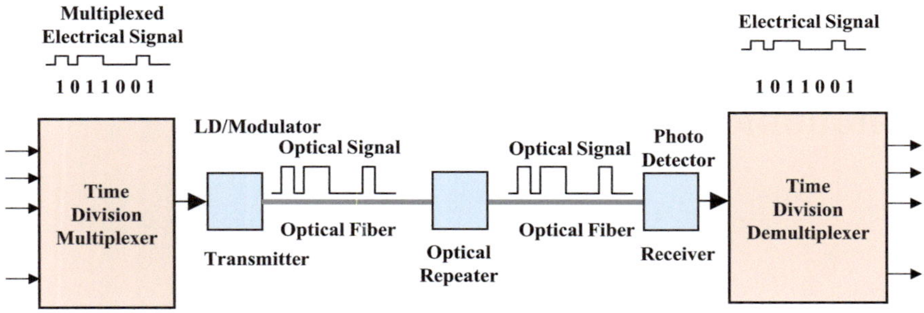

Fig. 2.1 Basic configuration of optical communication systems

receiver side, the transmitted optical signal is converted into an electrical signal by the photo6detector, and the TDM demultiplexing circuit demultiplexes to the low-speed electrical signals, transmitting the information to the destination.

Fundamental technologies such as optical fibers, semiconductor lasers, optical modulators, and photodetectors have been continuously innovated to meet the needs for higher speed and larger capacity in optical communication systems. The advent of optical communications began with the invention of the laser (Charles H. Townes, Nicolay G. Basov, Aleksandr M. Prokhorov, Nobel Prize in Physics 1964) and optical fiber (Charles Kao, Nobel Prize in Physics 2009). However, it is said that the practical beginning of optical communication was in 1970, when the continuous-wave operation of semiconductor lasers at room temperature (Izuo Hayashi and Morton B. Panish, AT&T Bell Lab, APL, p. 109, 1970) and low-loss optical fiber (20 dB/km at 632.8 nm) (F. P. Kapron et al., Corning, APL, p. 423, 1970) were simultaneously demonstrated.

Table 2.1 summarizes the evolution of each fundamental technology from various perspectives. In the 1970s, the focus of optical fiber technology was on controlling spatial modes and losses; in the 1980s, it was on controlling chromatic dispersion; and from the 1990s onwards, after the advent of optical amplifiers, controlling nonlinearity became the key issue, leading to the research and development of various types of optical fibers. Recently, future fiber technologies such as multi-mode and multi-core optical fibers are being studied.

Regarding advances in optical device technology, the main challenge for semiconductor lasers in the 1970s and the 1980s was controlling the oscillation wavelength range to match the minimum loss wavelength of optical fibers (0.8, 1.3, and 1.55 μm wavelength bands). Subsequently, research and development focused on controlling the oscillation mode (single-mode lasers) and variable control of oscillation wavelengths, considering the wavelength dispersion of optical fibers. Due to interaction with chromatic dispersion of optical fiber, spectral broadening during modulation also became an issue, and from the 1980s to the 1990s, external modulators were developed as an alternative to direct modulation of semiconductor lasers. Regarding photodetectors, PIN photodiodes using semiconductors to match the wavelength

Table 2.1 Evolution of fundamental technologies in optical communication

	1970	1980	1990	2000	2010	2020
	Early Stage (Mbps)		Growth Stage (Gbps)	Maturity Stage (Tbps)		Next Generation (Pbps)
Optical fiber	Mode/Loss Control		Dispersion Control	Nonlinear Control		
	Multi-mode					Multi-mode/ Multi-core
	Single-mode	Dispersion Shift	Non-zero Dispersion Shift			
			Dispersion Ccompensation			
				Low-nonlinearity		
Laser	Wavelength Control		Mode control			
	0.8 μm	1.3 μm	1.5 μm single-mode	WavelengthTunable		
Photodiode	PIN photodiode	Avalanche Photodiode				
Multiplexing	TDM		WDM		QAM	SDM
Modulation	Intensity Modulation /Direct Detection	Direct Mdulation	External modulation	Differential Phase Modulation /Direct Detection	Vector Modulation/ Ccoherent Detection	
Repeater		Regenerator		Optical Amplifier		

bands used in optical communication systems and avalanche photodiodes offering excellent receiver sensitivity were developed.

The repeater technology shifted from regeneration repeaters to optical amplification repeaters. After the emergence of optical amplifiers in the 1990s, multiplexing methods transitioned from TDM to wavelength division multiplexing (WDM), leading to a rapid increase in system capacity. Recently, the transition from Intensity Modulation-Direct Detection (IM-DD) to coherent detection, where the phase of light can also be detected as in radio communications, has enabled further capacity increases through multiplexing in the quadrature amplitude domain (multi-level modulation). Space Division Multiplexing (SDM) is also being considered as a future technology.

In the following sections, an overview of each fundamental technology will be provided.

2.1 Optical Fiber

2.1.1 Multi-mode Fiber and Single-Mode Fiber

Optical fibers are the transmission medium in optical communication systems and thus represent the most important fundamental technology in the systems. Previously, wired communication systems utilized coaxial cables, but the advent of low-loss optical fibers significantly extended the transmission distance. The theory of optical fibers was proposed by Charles Kao in 1966, and after Corning Inc. realized optical fibers with a loss of 20 dB/km, considered low at that time, in 1970, the technology rapidly advanced.

As shown in Fig. 2.2, an optical fiber consists of a core with a high refractive index surrounded by cladding with an outer diameter of 125 μm. The difference in refractive index between the core and the cladding is approximately 0.3–1%, allowing light to be confined and propagated within the core. Although the propagation of light waves in optical fibers is described by wave equations, it is sometimes represented by a simple ray model, as shown in Fig. 2.2, for simplicity. In this model, only light that enters at an angle less than or equal to the critical angle for total internal reflection is confined and propagated within the core.

Optical fibers are mainly classified into two types: multi-mode fibers (MMFs) with a core diameter of about 50 μm and single-mode fibers (SMFs) with a core diameter of about 10 μm (see Fig. 2.3). In MMF, multiple propagation modes with different spatial light distributions exist within the core, each with different propagation speeds (modal dispersion), and the distribution of each mode changes over time (modal coupling). Modal coupling leads to temporal fluctuations in the received waveform, and modal dispersion results in the spreading of the received waveform, making MMF unsuitable for long-distance or high-speed signal transmission. MMF is further categorized into step-index (SI) MMF and graded-index (GI) MMF based on the

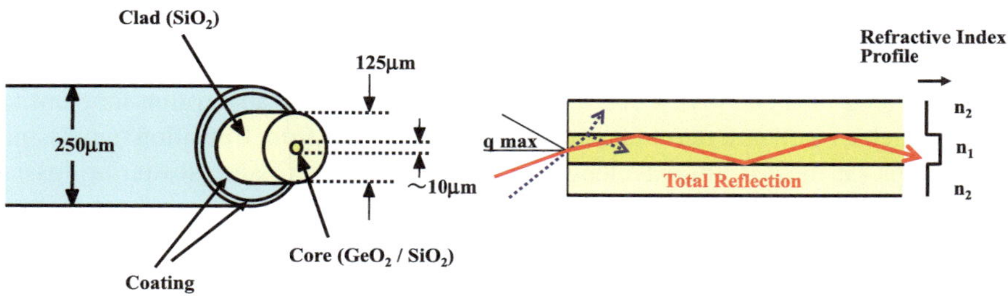

Fig. 2.2 Structure and guiding principle of single-mode fiber

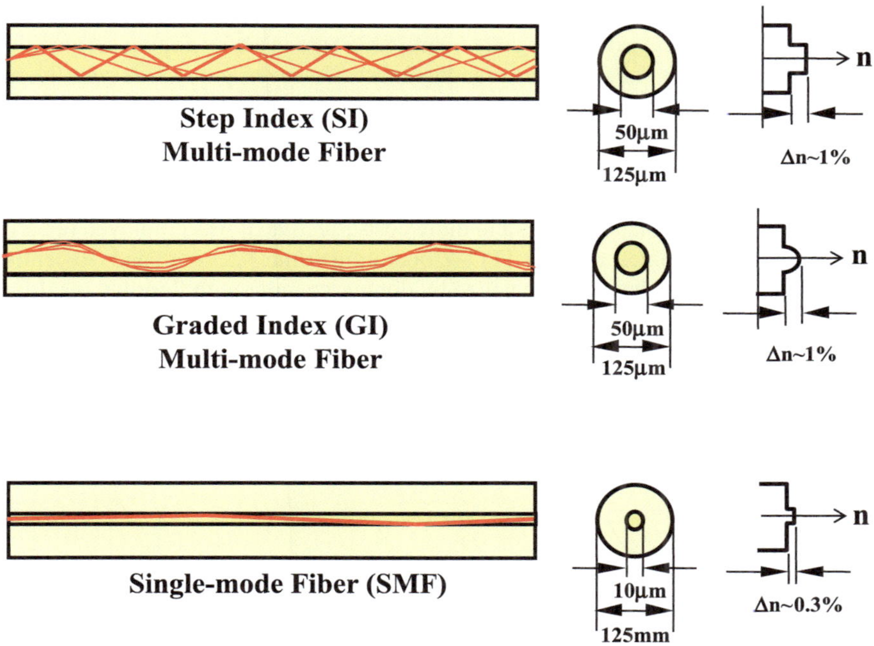

Fig. 2.3 Types of optical fibers

refractive index profile of the core, with the latter being designed to minimize modal dispersion. In contrast, SMF supports only a single propagation mode, eliminating modal dispersion.

2.1.2 Basic Characteristics of SMF

(1) Loss Characteristics

Figure 2.4 shows the wavelength dependence of optical fiber loss. The losses in optical fibers are categorized into absorption loss and scattering loss. Absorption losses include intrinsic losses due to ultraviolet and infrared absorption inherent to the material, as well as external factors such as impurities from transition metals and OH groups (at the 1.4 μm wavelength band). On the other hand, scattering losses include Rayleigh scattering due to the glass structure, as well as external factors like irregularities at the core-cladding interface and the inclusion of bubbles during manufacturing. Rayleigh scattering is inversely proportional to the fourth power of the wavelength of light. Initially, the 0.8 μm band was used for optical fiber communication, but as external losses were reduced and performance improved, the wavelength shifted to the 1.3 μm band. Currently, optical fibers with a loss of less than 0.2 dB/km in the 1.55 μm band, which is the lowest loss wavelength band, are commonly used. Standard single-mode fibers (SMF) are made by doping germanium (Ge) into the core to increase the refractive index. However, pure silica core fibers, where the core is made of pure SiO_2 without impurities and the cladding is doped with fluorine to reduce its refractive index, have also been developed to reduce losses. This type of optical fiber is extremely important in regeneration repeater systems that perform optical–electrical–optical conversion in repeaters, as it allows for extended repeater spacing and makes the system more cost-effective. It was employed in early optical submarine cables, such as TPC-4, operating at the 1.55 μm band. Recently, further loss reduction has been achieved by doping the core with fluorine, resulting in ultra-low-loss fibers with 0.1397 dB/km [1].

Loss reduction in optical fibers was crucial for extending repeater spacing in regeneration repeater systems. In optical amplifier repeater systems, as will be discussed

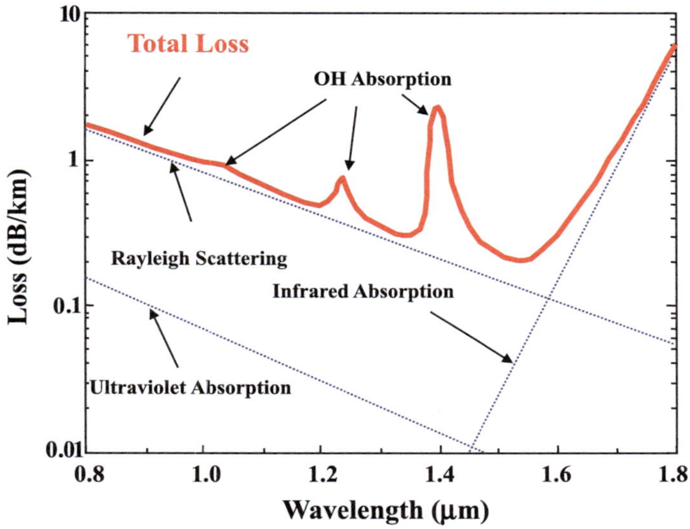

Fig. 2.4 Wavelength-dependent loss of optical fiber

later, it remains vital for suppressing nonlinear effects in the fiber, as it allows for a reduction in input power even with the same repeater spacing. Therefore, it continues to be an important topic in current research and development.

(2) Chromatic Dispersion

Another characteristic of optical fibers is their chromatic dispersion characteristics. Figure 2.5 shows the wavelength dependence of chromatic dispersion for various types of optical fibers. Chromatic dispersion is determined by the material dispersion inherent to the glass and the structural dispersion (waveguide dispersion) determined by the core shape, among other factors. The unit of chromatic dispersion, D, is expressed in (ps/nm/km). For instance, it represents the time difference in picoseconds between two light signals separated by 1 nm after transmitting through 1 km of optical fiber.

In long-distance or high-speed transmission, chromatic dispersion causes temporal broadening of the received optical signal, making it essential to narrow the spectral width of the modulated light or to minimize the fiber's chromatic dispersion to reduce waveform distortion. In standard SMF, the dispersion value D is approximately 0 ps/nm/km at the 1.3 μm band and about 18 ps/nm/km at the 1.55 μm band. In communication at the 1.3 μm wavelength band, the chromatic dispersion is close to zero and the spectral broadening of the light source is not an issue, allowing for the use of multi-mode light sources such as Fabry–Perot lasers without problems. However, in long-distance or high-speed communication at 1.55 μm band with the lower loss, the spectral broadening of the light source becomes a significant issue due to chromatic dispersion.

To mitigate the effects of chromatic dispersion, dispersion-shifted fibers (DSFs) were developed by altering the refractive index distribution of the fiber core (e.g.,

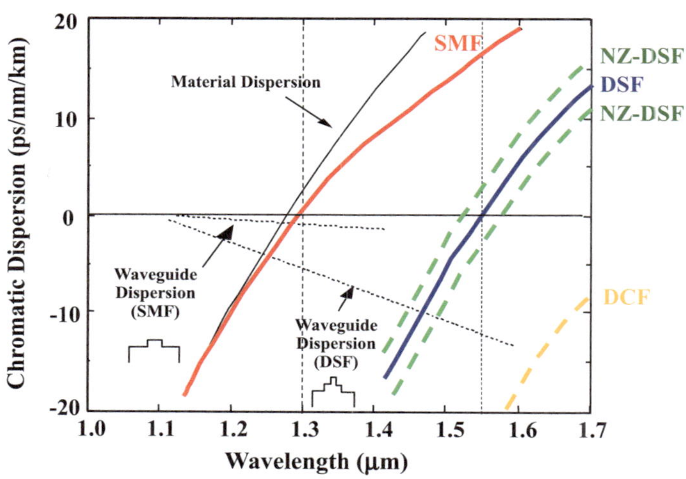

Fig. 2.5 Wavelength dependence of chromatic dispersion of optical fiber

stepped profile) to create a negative structural dispersion that cancels out the material dispersion, shifting the zero-dispersion wavelength to the 1.55 μm band. This allowed for long-distance, high-speed communication at the 1.55 μm band. However, as will be discussed later, in wavelength division multiplexing (WDM) systems, zero dispersion can exacerbate signal degradation due to fiber nonlinearity. Consequently, non-zero DSF (NZ-DSF) was developed with dispersion values of ± 2–4 ps/nm/km at the 1.55 μm band.

In optical transmission systems, the dispersion value *D* of the fiber accumulates with distance *L*, requires significant dispersion compensation at the receiver side. To avoid this, the dispersion management scheme (see Chap. 6) was introduced in long-haul transmission systems. In this scheme, dispersion compensation fibers (DCFs) are periodically inserted in the transmission line, such that the cumulative dispersion value of DCFs becomes -DL. For this purpose, DCFs with negative dispersion values ranging from − 20 to − 100 ps/km/nm in the 1.55 μm band have also been developed. On the other hand, in the digital coherent detection systems (see Chap. 8), digital signal processing (DSP) in the receiver can compensate for the large accumulated dispersion in transoceanic submarine cable systems, which can remove the dispersion management in the transmission line.

Another critical characteristic related to chromatic dispersion in optical fibers is the dispersion slope, which represents the wavelength dependence of chromatic dispersion. The unit of dispersion slope is (ps/nm^2/km). In WDM systems, the dispersion slope causes different chromatic dispersion values for the center wavelength and other wavelengths, making it difficult to achieve uniform transmission characteristics across a broad range of wavelengths. To suppress this and achieve transmission characteristics with minimal wavelength dependence, optical fibers with low dispersion slopes or dispersion-flattened fibers are crucial.

(3) Polarization Mode Dispersion

The phenomenon where light with different polarizations propagates at different speeds through an optical fiber is called Polarization Mode Dispersion (PMD), with the unit expressed as ps/$\sqrt{}$km (see Fig. 2.6). This occurs due to deviations from the perfect circular core diameter, internal stress and strain generated during manufacturing, or uneven external stress and strain. A propagation delay time difference between polarizations caused by PMD, known as Differential Group Delay (DGD), varies due to disturbances such as temperature changes and vibrations. Using optical fibers with high PMD can result in temporal fluctuations and waveform broadening of the received signal, which limits high-speed and long-distance transmission. While PMD values in early optical fibers were relatively large, it is now common for PMD link-design value to be less than 0.05 ps/$\sqrt{}$km.

(4) Bending Loss

The characteristic where light is not sufficiently confined to the core of the optical fiber when the fiber is bent, resulting in loss, is called macro-bending loss. Macro-bending loss can be mitigated by increasing the refractive index difference between

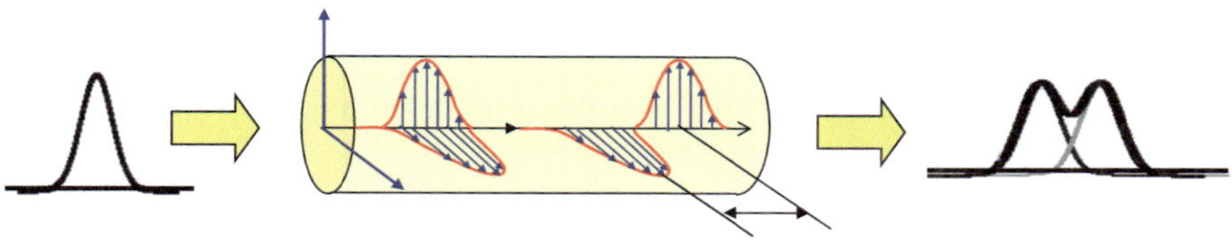

Fig. 2.6 Polarization mode dispersion of optical fiber

the core and the cladding, which enhances light confinement within the core. On the other hand, to suppress fiber nonlinearity, it is necessary to either enlarge the core diameter or reduce the refractive index difference between the core and cladding, which weakens light confinement within the core and tends to increase macro-bending loss. Therefore, it is important to design optical fibers considering both aspects. Additionally, macro-bending loss exhibits wavelength dependence, with longer wavelengths being more sensitive to bending and resulting in higher bending loss. Therefore, special attention is required for L-band systems using the 1.6 μm wavelength range. Another characteristic related to the bending is micro-bending loss, which occurs due to irregular pressure from the surrounding environment and affects light confinement. It depends on the characteristics of the coating material besides the waveguide design of the optical fiber.

2.1.3 Nonlinear Characteristics of SMF

Nonlinear phenomena in optical fibers refer to the changes in the fiber's properties (such as refractive index) in response to the intensity of the light. Since the absolute value of these nonlinear effects is small, they did not pose a significant problem in regeneration repeater systems where light is converted to electrical signals and regenerated approximately every 100 km. However, with the advent of optical amplifiers, where light emitted from the transmitter is attenuated by the optical fiber and amplified by the optical amplifier repeatedly, allowing transmission over distances ranging from hundreds to thousands of kilometers, even small absolute values of fiber nonlinearity in silica glass accumulate with distance and become significant problems.

Generally, optical nonlinear effects are represented by [2]:

$$P = \chi^{(1)}E + \chi^{(2)}E^2 + \chi^{(3)}E^3 + \cdots \tag{2.1.1}$$

where P is the total polarization, E is the optical electric field, $\chi^{(1)}$ represents linear susceptibility such as refractive index or absorption coefficient, and $\chi^{(i)}$ (for $i \geq 2$) represents nonlinear susceptibility. Since $\chi^{(2)}$ vanishes in an isotropic material such as silica glass, optical fibers do not normally exhibit second-order nonlinear effects.

The main optical nonlinear effects in optical fibers include:

1. Self-phase modulation (including cross-phase modulation) due to the third-order nonlinear effect of nonlinear refractive index.
2. Four-wave mixing due to the third-order nonlinear effect.
3. Stimulated Raman Scattering due to fiber-induced scattering.
4. Stimulated Brillouin Scattering due to fiber-induced scattering.

The following sections provide an overview of each optical nonlinear effect.

(1) Self-phase Modulation/Cross-Phase Modulation

The optical nonlinear effect (Kerr effect) in optical fibers refers to the phenomenon where the refractive index of the glass changes slightly depending on the light power (energy density), and is represented as follows:

$$n = n_0 + n_2 \cdot |E|^2 \tag{2.1.2}$$

$$|E|^2 = P_O / A_{\mathrm{eff}} \tag{2.1.3}$$

where n is the refractive index, n_0 is the linear refractive index, n_2 is the nonlinear refractive index, P_O is the optical power, and A_{eff} is the effective area of the core. The nonlinear coefficient γ, which represents the magnitude of the nonlinearity in the optical fiber, is given by:

$$\gamma = \frac{2\pi}{\lambda} \cdot \frac{n_2}{A_{\mathrm{eff}}} \tag{2.1.4}$$

where λ is the wavelength of the light.

Self-phase modulation (SPM) is a phase modulation that occurs due to changes in optical intensity. In the case of optical pulses shown in Fig. 2.7, the refractive index increases in the central part of optical pulse with high power due to the Kerr effect, resulting in a larger phase delay at that location. When the phase change is differentiated, it is converted into an instantaneous frequency change. In this case, a lower frequency component (longer wavelength component) is generated at the rising edge of the signal, while a higher frequency component (shorter wavelength component) is produced at the falling edge, resulting in frequency chirp. This frequency chirp caused by SPM becomes larger as the rise and fall of the signal become steeper in higher-speed signals.

Cross-phase modulation (XPM) is a phase modulation that occurs when an optical signal at a different wavelength overlaps temporally with the signal, causing changes in the intensity of the combined optical signal. When chromatic dispersion is small or the wavelength spacing between the signals is narrow, i.e., dense wavelength division multiplexing (DWDM), the relative positional relationship of adjacent wavelength optical pulses is maintained over long distances, increasing the influence of XPM.

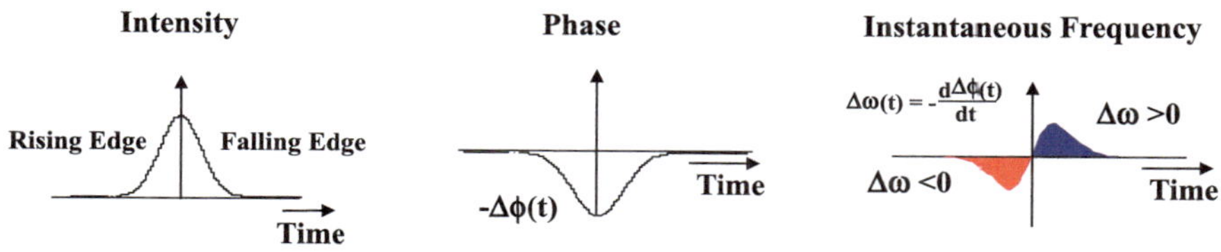

Fig. 2.7 Self-phase modulation

(2) Four-Wave Mixing

Four-wave mixing (FWM) is a nonlinear effect arising from third-order optical nonlinearity. When two optical signals of different wavelengths propagate through a fiber, a refractive index modulation occurs due to the beat signal (frequency: $f_2 - f_1$) between the two wavelengths. As shown in Fig. 2.8, two new frequency components are generated on either side of the two original wavelengths. In wavelength division multiplexing (WDM) systems with equal frequency spacing, FWM light can cause crosstalk, degrading signal quality. Conditions favorable for FWM include small chromatic dispersion, narrow wavelength spacing, and aligned polarization. In such cases, the involved optical signals move at nearly the same speed (phase matching), leading to an increased interaction length and a greater impact of FWM.

In single-channel transmission systems, signals are transmitted near the zero-dispersion wavelength. However, in WDM systems, transmitting near the zero-dispersion wavelength generates many new FWM components, increasing crosstalk and noise in the signal light. Therefore, in WDM systems, to avoid phase matching, fibers with a certain degree of dispersion such as Non-Zero-Dispersion-Shifted Fibers (NZ-DSFs) or Standard Single-Mode Fibers (SMFs) are used as transmission fibers, and the accumulated dispersion is compensated periodically in the transmission line (see Chap. 6). In recent digital coherent detection systems (see Chap. 8), chromatic dispersion can be compensated by digital signal processing at the receiver, allowing the use of fibers with the large chromatic dispersion as standard SMF, minimizing the effects of FWM and XPM.

(3) Stimulated Raman Scattering

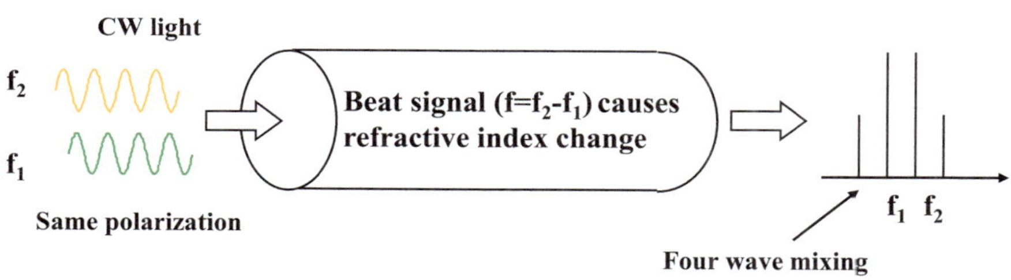

Fig. 2.8 Four-wave mixing

When strong excitation light (pump light) is incident on silica glass, scattering light (Stokes light) with a frequency slightly lower than that of the excitation light is generated due to the interaction between the lattice vibrations (optical vibrations) of the silica crystal and the excitation light. When very strong excitation light propagates through an optical fiber, a large portion of the pump light's energy is converted into Stokes waves, leading to Stimulated Raman Scattering (SRS). The scattered light reaches its peak gain at a frequency determined by the medium, and in the case of pure silica, this peak gain occurs at a frequency 1.32 THz lower than that of the pump light, approximately 100 nm longer wavelength in the 1.5 μm band. For example, if a pump light at 1.45 μm and a signal light at 1.55 μm about 100 nm apart are simultaneously launched into the optical fiber, the signal light is amplified through SRS (Raman amplification), making it possible to turn the transmission path itself into an optical amplifier.

(4) Stimulated Brillouin Scattering

Stimulated Brillouin Scattering (SBS) is a phenomenon where light interacts with acoustic waves in a medium, resulting in scattering with a slight shift in frequency. In optical fibers, the excitation light generates acoustic modes of lattice vibrations due to the electrostriction effect, creating a periodic modulation of the refractive index, a diffraction grating. The pump light is scattered due to this diffraction grating, which moves backwards at the speed of sound, causing the scattered light to shift to the lower-frequency side due to the Doppler effect. For the 1.55 μm band, the frequency of the scattered light shifts about 11 GHz to the lower-frequency side compared to the incident light. When light signals that exceed the SBS threshold are launched into an optical fiber, the incident light is reflected back to the input end due to Brillouin scattering, preventing an increase in the input light power. SBS becomes more pronounced with higher coherence of the incident light, so broadening the spectral width of the signal light, for example by adding phase modulation, can suppress it.

2.1.4 Optical Fibers in Trans-Pacific Submarine Cable Systems

Having discussed the basic overview of optical fibers, it is noted that in actual Trans-Pacific optical submarine cable systems, improvements have continuously been made considering factors such as loss, chromatic dispersion, and fiber nonlinearity according to the times. Table 2.2 shows the types of optical fibers used in Trans-Pacific optical submarine cable systems.

Table 2.2 Optical fibers used in Trans-Pacific optical submarine cable systems

TPC-3 (1989)	Standard SMF with zero dispersion at 1.3 μm
TPC-4 (1992)	Pure silica core SMF with low loss at 1.55 μm
TPC-5 (1995)	DSF with zero dispersion at 1.55 μm
China-US (2000)	NZ-DSF with non-zero dispersion at 1.55 μm and SMF for dispersion compensation
Japan-US (2001)	Effective area enlarged low-nonlinear NZ-DSF, low dispersion slope NZ-DSF and SMF for dispersion compensation
UNITY (2010)	Effective area enlarged low-nonlinear SMF and DCF for dispersion and dispersion slope compensation
FASTER (2016)	Effective area enlarged low-nonlinear SMF

2.2 Semiconductor Laser for Long-Haul Transmission Systems

2.2.1 Basic Structures of a Semiconductor Laser

Semiconductor lasers are the light sources for optical communication systems, and the research and development of compact and reliable semiconductor lasers have led to the current state of optical communication. Laser oscillation, which involves a significant increase of optical output power once the drive current exceeds the threshold, requires both light emission and optical resonance. Indirect transition semiconductors such as Si and Ge are unsuitable for light-emitting devices, so compound semiconductors with direct transitions, such as AlGaAs/GaAs and InGaAsP/InP, are used. Initially, GaAs-based AlGaAs/GaAs lasers were developed as semiconductor lasers for optical fiber communication. The emission wavelength of these lasers is in the 0.8–0.9 μm range. In the early 1970s, the loss in optical fibers was minimal around this wavelength and was approximately 20 dB/km, and it matched the high-sensitivity wavelength range of Si avalanche photodiodes (Si-APD) developed as detectors.

Later, as it was predicted that optical fiber transmission losses would be lower at wavelengths longer than 1 μm, InGaAsP/InP lasers that oscillate in the 1 μm range became mainstream for long-distance communication. Figure 2.9a shows the basic structure of such lasers. As shown in Fig. 2.9b, the InGaAsP emission layer has a smaller energy bandgap compared to the InP cladding layer, which allows holes injected from p-type InP and electrons from n-type InP to be confined and recombine efficiently to produce light. Furthermore, as shown in Fig. 2.9c, the refractive index of InGaAsP is higher than that of InP, which confines the light in InGaAsP.

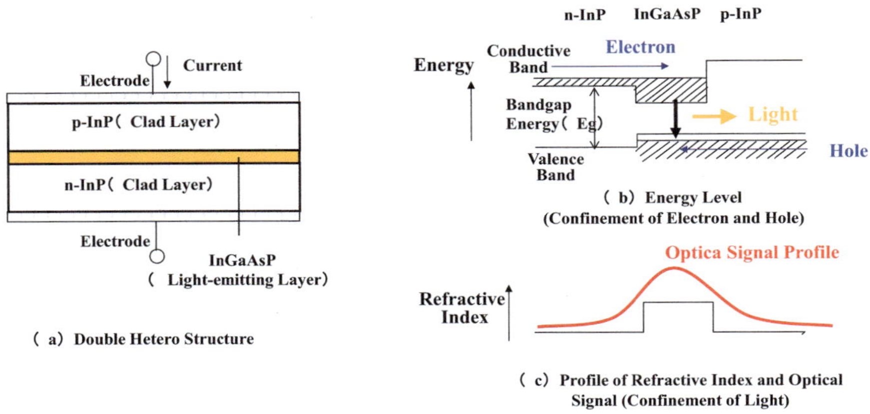

Fig. 2.9 Basic structure, energy level, and refractive index profile of semiconductor laser

This sandwich structure, known as a double heterostructure, is the basic structure of semiconductor lasers.

2.2.2 1.3 μm Fabry–Perot Lasers

For long-distance transmission, it is advantageous to have as low transmission loss as possible. Initially, semiconductor lasers that operate in the 1.3 μm band, where optical fiber chromatic dispersion becomes zero, were developed. At that time, optical fiber loss was minimal in the 1.3 μm band, approximately 0.5 dB/km. These lasers utilized a Fabry–Perot (FP) resonator with a pair of parallel reflecting mirrors and were multi-wavelength lasers that oscillated at multiple wavelengths around the wavelength where the emission was strongest. Improvements for achieving stable longitudinal multi-mode operation and reducing operating current enhanced their performance. Long-duration continuous operation tests demonstrated high reliability, leading to their introduction in the optical submarine repeaters of the third Trans-Pacific cable system (TPC-3), the first optical submarine cable system discussed later.

2.2.3 1.5 μm Single-Mode Lasers

As the optical fiber's loss reduction continued, the wavelength of minimum loss also shifted to longer wavelengths. It was predicted that the minimum loss would be around 0.2 dB/km at approximately 1.55 μm, and this was achieved in 1979 with the development of ultra-low-loss optical fibers. Therefore, semiconductor lasers operating at 1.55 μm were required. However, while the 1.55 μm band has the minimum optical fiber loss, it is characterized by significant chromatic dispersion (the property where the speed of light differs with wavelength, see Sect. 2.1). Using multi-wavelength laser sources similar to those at 1.3 μm would result in the spreading of optical pulses due to the fiber's chromatic dispersion. Consequently, semiconductor lasers that operate at only one wavelength were needed, leading to intense global research competition starting from late 1970s.

The leading candidate for this was the Distributed Bragg Reflector (DBR) or Distributed Feedback (DFB) Laser, which incorporates periodic diffraction gratings with a period equal to half the wavelength inside the semiconductor within the laser.

Figure 2.10a, b show the structures of these lasers. This periodic structure is known as a diffraction grating, which exhibits strong reflective characteristics at a wavelength that is twice the period of the grating called the Bragg wavelength for first-order diffraction gratings. Initially, the DFB laser, which integrates the light-emitting region with the reflecting region as shown in Fig. 2.10b, was put into practical use. A DFB laser fundamentally operates at two wavelengths slightly away from the Bragg wavelength, and it is not a single-wavelength laser. However, if one of the anti-reflection coatings is removed or replaced with a high-reflection coating, the structure becomes asymmetric, and only one of the two wavelengths might be emitted. The probability of stable single-wavelength emission is not very high. To achieve stable single-wavelength operation precisely at the Bragg wavelength, it is necessary to shift the phase of the grating's periodic structure at the center [3]. This results in a structure where the phase is shifted by one-quarter of the wavelength, leading to the term 1/4 wavelength shift DFB laser. Figure 2.11 illustrates an example of a 1/4 wavelength shift DFB laser with an embedded stripe structure.

In place of the anti-reflection coating shown in Fig. 2.10b, the output end is designed with an embedded structure to effectively achieve anti-reflection. Figure 2.12 displays the spectra at 0.9 times the threshold current (I_{th}), clearly illustrating the spectral characteristics of the DFB laser.

Figure 2.12a shows the spectrum of a 1/4 wavelength shift DFB laser made using the aforementioned method, which exhibits a sharp single peak exactly at the Bragg wavelength (λ_0). In contrast, Fig. 2.12b shows the spectrum of a DFB laser with a uniform diffraction grating, where two sharp peaks appear on either side of the Bragg wavelength. Although a uniform diffraction grating DFB laser can sometimes achieve single-wavelength emission if the end faces act as reflective ends due to asymmetry, the probability of single-wavelength operation is relatively low. On the other hand, the 1/4-wavelength-shift DFB laser, with anti-reflective ends on both sides, theoretically provides 100% single-wavelength operation, and high manufacturing yields have been demonstrated.

For commercial deployment, it is essential to confirm whether single-wavelength operation remains stable over the long term. Figure 2.13 shows examples of long-term lifetime test results, including spectral characteristics. The 1/4 wavelength shift DFB laser was continuously operated at 50 °C for 3000 h. The measured spectra before and after continuous operation with increased drive current are shown. The lower section (linear scale) shows the spectrum at $I = 0.9I_{th}$, and the upper section (log scale with 10 dB/div) shows the spectrum at $I = I_{th} + 5$ mA with a 1 GHz modulation current I_m

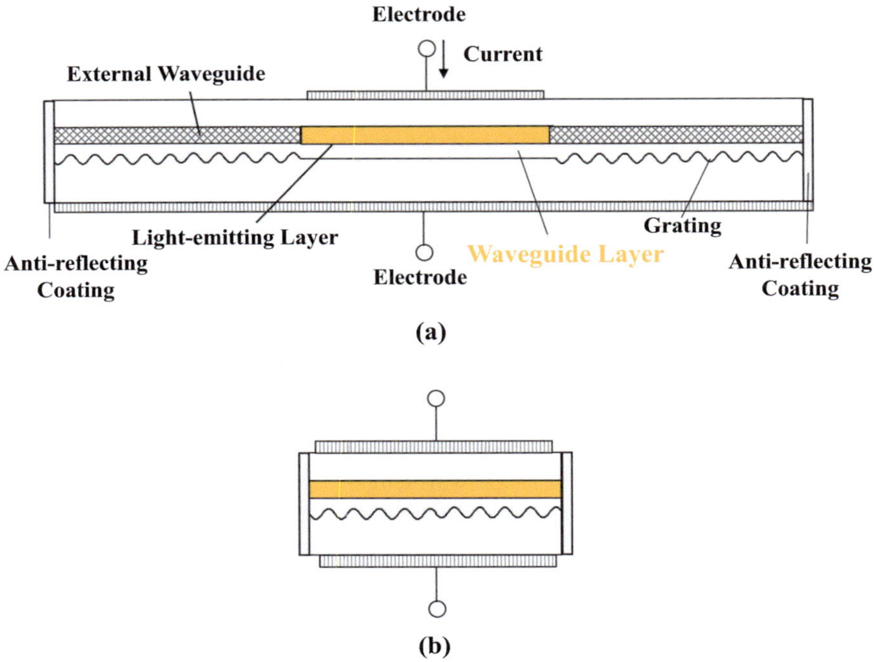

Fig. 2.10 Basic structure of **a** distributed Bragg reflector laser and **b** distributed feedback laser

= 30–40 $mA_{p\text{-}p}$ superimposed for high-speed direct modulation. Despite the increase in operating current due to the potential degradation, the spectrum remains stable due to the inherent single-wavelength characteristics, confirming its reliability for use in submarine repeaters.

Figure 2.14 shows the results of a lifetime test using 230 units of the 1/4 wavelength shift DFB lasers with the same specifications as those for commercial use. The histogram displays the change in the Side Mode Suppression Ratio (SMSR) with direct modulation at 565 Mb/s before and after 6500 h operation at temperatures of 10, 40, 50, and 60 °C. There was no significant change in the SMSR distribution before and after the lifetime test, confirming that there are no issues with implementing these lasers in commercial repeaters.

These single-mode DFB lasers were used in the repeaters of the fourth Trans-Pacific cable (TPC-4), significantly contributing to the realization of 1.55 μm band optical submarine cable systems.

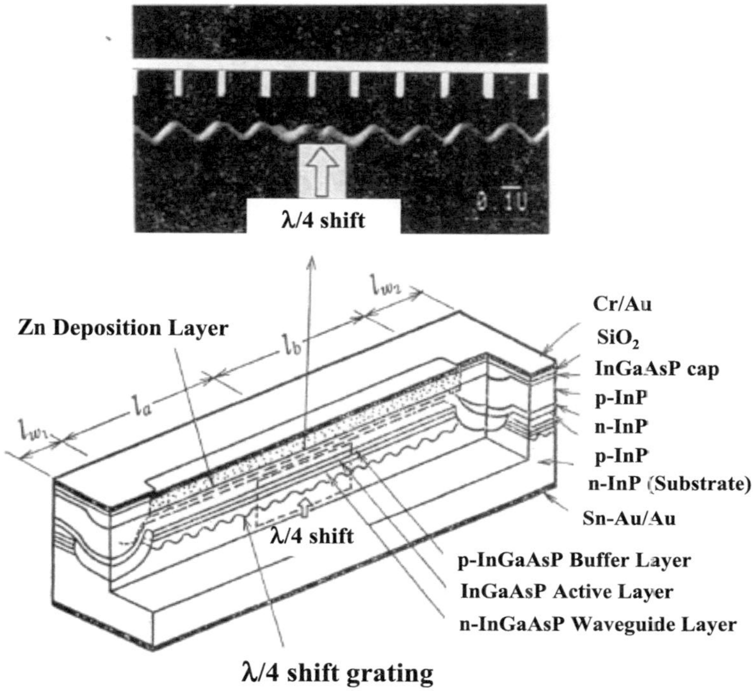

Fig. 2.11 Structural schematic of 1.55 μm InGaAsP/InP $\lambda/4$ shift DFB laser and cross-sectional photograph around $\lambda/4$ shift

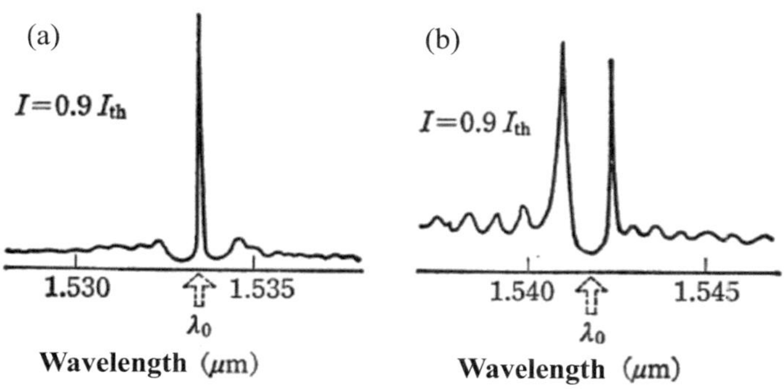

Fig. 2.12 Spectra at 0.9 times the threshold current: **a** 1/4 wavelength shift DFB laser, **b** DFB laser with a uniform grating

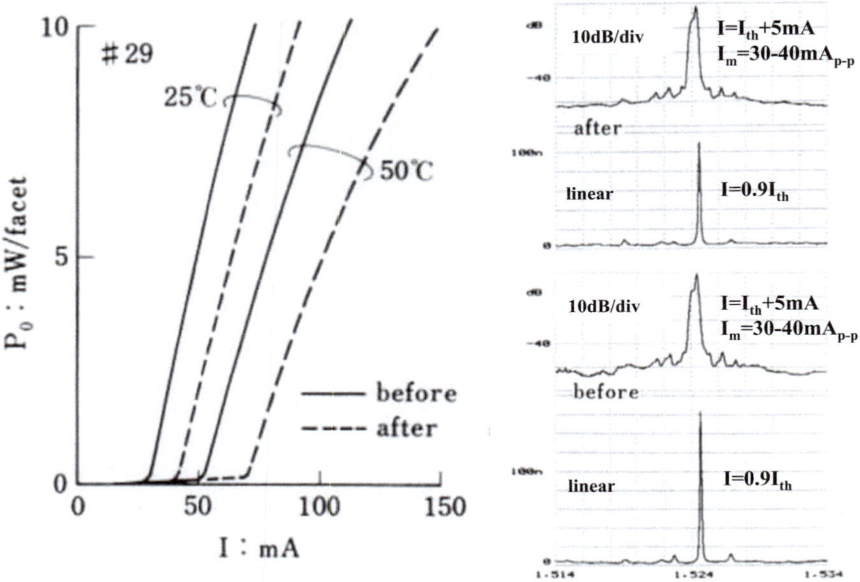

Fig. 2.13 Optical output power characteristics and optical spectra before and after long-term lifetime test

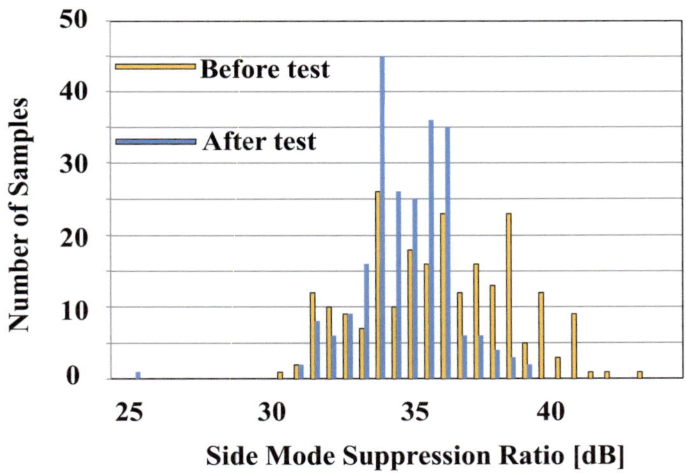

Fig. 2.14 Results of a lifetime test using 230 units of 1/4 wavelength shift DFB lasers

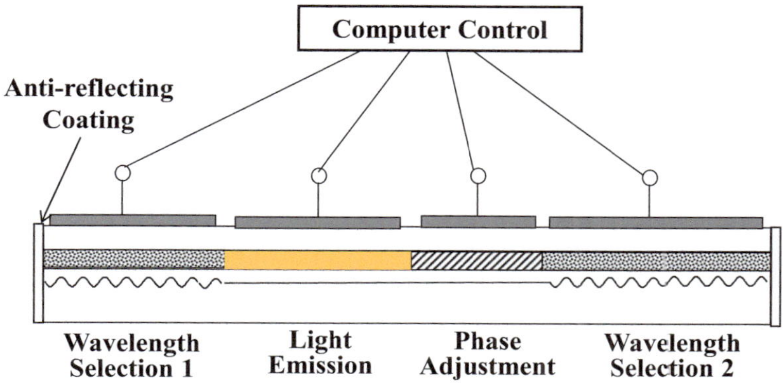

Fig. 2.15 Tunable laser in PIC

2.2.4 Wavelength Tunable Lasers

Subsequently, as wavelength division multiplexing systems became mainstream, wavelength-tunable lasers became widely used. As shown in Fig. 2.15, the basic configuration consists of an emission section, a phase adjustment section, and wavelength selection sections 1 and 2. Through computer control via their respective electrodes, these lasers can select any wavelength over a fairly wide range, making them one type of photonic integrated circuit (PIC). Additionally, there has been progress in developing devices with integrated functions such as modulation and optical amplification, as well as hybrid photonic ICs combining InGaAsP/InP compound semiconductors with silicon.

Another type of wavelength-tunable laser is a tunable external-cavity laser (ECL). The basic structure of the tunable ECL is shown in Fig. 2.16. The tunable ECL typically uses a diffraction grating as the wavelength-selective element in the external cavity. The diffracted beam provides optical feedback to the laser diode chip, which has an anti-reflection (AR) coating on the right side. The emission wavelength can be tuned by rotating the diffraction grating. Since the cavity length is much longer than the DFB laser, the linewidth of the tunable ECL is much narrower than DFB lasers. The typical linewidth of DFB lasers is on the order of several MHz to several tens of MHz, whereas that of tunable ECLs is below 100 kHz.

2.3 Optical Modulator

The process of varying the phase or intensity of light according to an external signal is known as optical modulation. This technique converts digital electrical signal information into optical signals. There are two main types of optical modulation

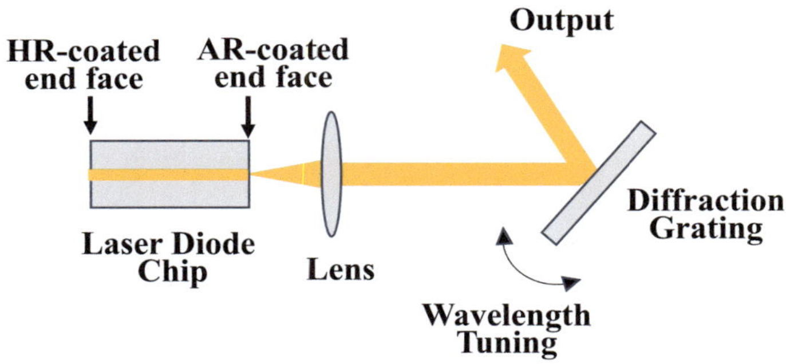

Fig. 2.16 Basic structure of a tunable external-cavity laser

methods: direct modulation and external modulation. In direct modulation, information is directly encoded by applying a modulated current to the laser. This method is the simplest but has limitations in response speed and causes significant fluctuations in the semiconductor's refractive index during the rise and fall times of the optical signal, leading to dynamic wavelength variations (chirping). Such dynamic wavelength variations can cause the optical waveform broadening or narrowing at the receiving end due to the chromatic dispersion of the optical fiber, degrading transmission performance. On the other hand, external modulation involves operating the light source at a constant output and modulating the signal using an external optical modulator. This method can suppress dynamic wavelength chirping, making it essential for high-speed optical transmission systems and optical amplification systems. Optical intensity modulators are categorized into two groups, i.e., a modulator based on Mach–Zehnder interferometer using electro-optic effect of dielectric material such as lithium niobate ($LiNbO_3$) and a modulator using electro-absorption (EA) effect in the semiconductor such as InGaAsP. A $LiNbO_3$ modulator can realize both an intensity modulator for IM-DD systems and an intensity and phase modulator for coherent communication systems. Current long-haul transmission systems mainly use $LiNbO_3$ modulators. On the other hand, EA modulators can be monolithically integrated with laser diodes and are suitable for compact modulator-integrated light sources for short-haul or middle-haul IM-DD transmission systems. The recent characteristics of these two types of optical modulators are summarized in Table 2.3. The details of these modulators are described in the following sections.

Table 2.3 Achieved characteristics of the recent optical modulators

	Bandwidth (GHz)	Driving voltage (V)	Main application
$LiNbO_3$ modulator	110	2.4	Long-haul IM-DD systems and coherent systems
EA modulator	85	0.8	Short and middle haul IM-DD systems

2.3.1 Optical Modulators Using the Electro-Optic Effect

In optical modulators, digital optical signals are generally generated by modulating the refractive index or absorption coefficient of a semiconductor or dielectric material using an external electrical signal. The change in refractive index due to the applied voltage is known as the electro-optic effect, and when the refractive index change is proportional to the electric field, it is referred to as the Pockels effect. For electro-optic crystals, $LiNbO_3$, which has a relatively large electro-optic coefficient and is stable in air, is commonly used.

Figure 2.17 illustrates the operational principle of an optical modulator using the Pockels effect. When a voltage V is applied to the electro-optic crystal, an electric field $E(= V/d)$ is generated inside the crystal with thickness d, causing the refractive index to change from n to $n + \Delta n$ in proportion to the electric field. When light with wavelength λ and a specific polarization direction is incident on this crystal, the phase velocity of the light wave propagating through the crystal of length L also changes, resulting in a phase shift of the output light given by:

$$\phi = 2\pi \Delta nL/\lambda \tag{2.3.1}$$

The amount of phase shift depends on the polarization direction of the incident light and the material. In this example, a bulk optical modulator directly applied the electric field to the crystal is shown. However, to more effectively apply the electric field to the optical wave, it is necessary to construct a waveguide-type optical modulator that confines the incident light within a three-dimensional optical waveguide.

For optical communication systems using Intensity Modulation-Direct Detection (IM-DD), it is necessary to modulate the intensity of the light. Figure 2.18 illustrates a Mach–Zehnder interferometer used to convert phase modulation into intensity modulation. When light is incident on the Mach–Zehnder interferometer, the incident light is split into two optical waveguides and then recombined after propagating through these waveguides. When no voltage is applied to the waveguides, the phases

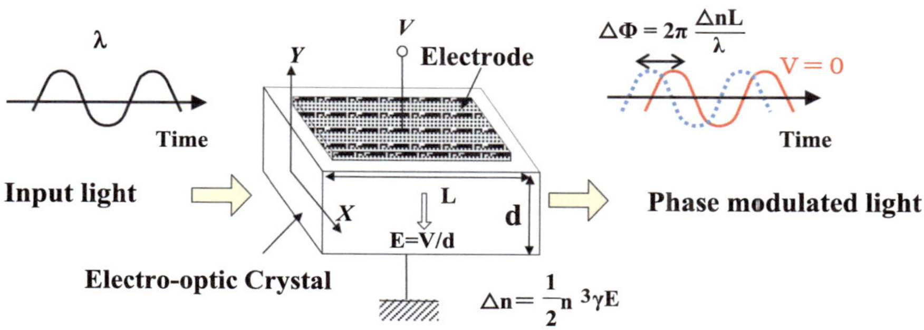

Fig. 2.17 Pockels effect

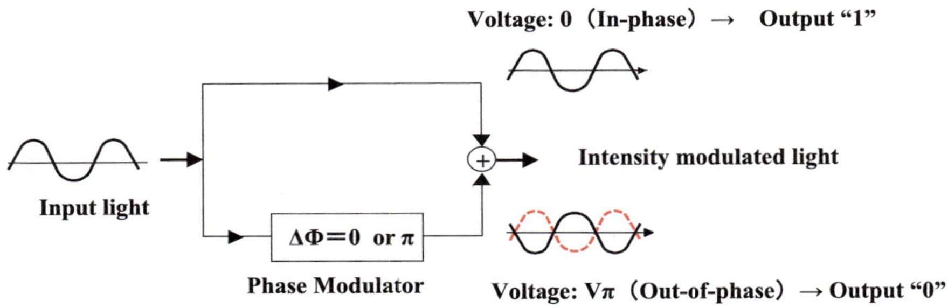

Fig. 2.18 Mach–Zehnder type optical modulator

of the two optical waves that recombine are the same, so the incident light is simply recombined and output as is ("ON"). However, when a voltage is applied to one of the waveguides, causing a phase shift of π, the phases of the two waves become out of phase with each other, resulting in destructive interference and no output light ("OFF"). Thus, a digital voltage signal is used to convert phase modulation into intensity-modulated digital optical signals.

(1) Mach–Zehnder Optical Modulator with Push-Pull Operation

The Mach–Zehnder optical modulator has been extensively studied as one of the most fundamental waveguide-type optical modulators. As previously mentioned, its operation principle is based on splitting the light into two optical waveguides and modulating the phase of each waveguide. As shown in Fig. 2.19, by incorporating modulation sections into both split waveguides and applying phase shifts that are out of phase with each other, a push-pull operation can be achieved. This approach effectively doubles the phase difference for the same voltage, allowing for the realization of a low-voltage-driven optical modulator.

The waveguide length of optical modulation elements using the electro-optic effect is typically on the order of centimeters. In a conventional electrode configuration with a lumped constant, the frequency characteristics of the optical modulator are determined by the CR time constant. Therefore, if the element length is long, the capacitance C increases, and the CR time constant limits the operating speed. To

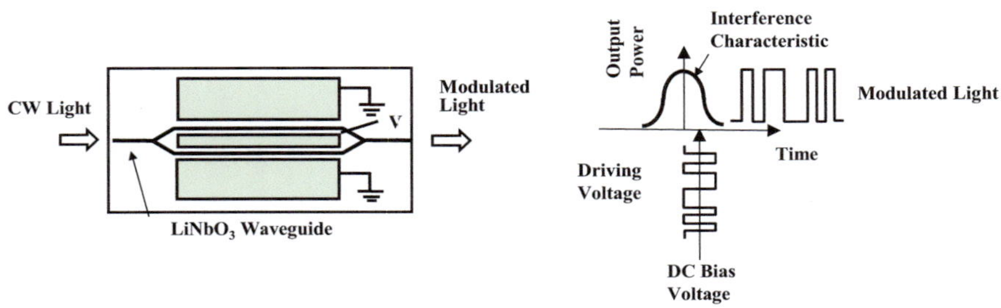

Fig. 2.19 Push-pull operation in Mach–Zehnder optical modulator

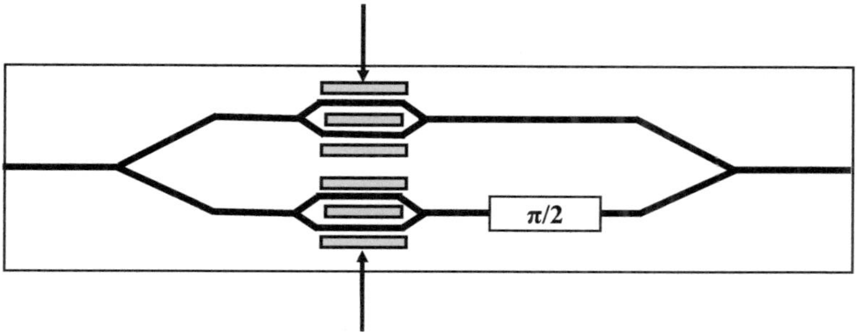

Fig. 2.20 Vector modulator for coherent communication systems

overcome this speed limitation, traveling-wave optical modulators have been developed where the propagation direction of light is matched with the direction of the electrical field. In Fig. 2.19, a voltage is applied from the left side, and a 50 Ω termination is provided on the right side. By designing the optical waveguide and electrode to match the effective refractive index of the light and the effective refractive index of the microwave as closely as possible, the frequency bandwidth is significantly expanded, enabling the development of high-speed modulators.

(2) Vector Modulators

So far, we have discussed intensity modulators. However, in coherent communication systems, which will be covered in Chap. 8, it is necessary to apply modulation independently to both intensity and phase. Optical modulators that modulate both intensity and phase are referred to as vector modulators. A representative example is the dual-parallel Mach–Zehnder modulator, shown in Fig. 2.20. This structure is realized by replacing the two arms of the Mach–Zehnder modulator with a Mach–Zehnder interferometer instead of a single waveguide. When light with a constant amplitude is incident on the optical modulator shown in Fig. 2.20, the upper arm generates modulation signals for the in-phase component, while the lower arm generates modulation signals for quadrature phase component with a phase shift of $\pi/2$. The combined signal from both arms is then output.

2.3.2 *Electro-Absorption Modulators*

Electro-optic optical modulators change the refractive index through an electric field to perform phase modulation or intensity modulation. In contrast, in semiconductors, it is possible to perform intensity modulation by controlling the absorption coefficient. As a general property of semiconductors, the absorption coefficient is high for light waves with wavelengths shorter than the absorption edge wavelength (λ_g) corresponding to the energy of the semiconductor's bandgap (E_g), and it is low for longer wavelengths. Franz and Keldysh demonstrated that when a voltage is applied to a semiconductor, the absorption edge wavelength shifts toward longer wavelengths in proportion to the square of the electric field, resulting in a significant increase in the absorption coefficient for longer wavelengths of light. This effect is known as the Franz–Keldysh effect, and in multi-quantum well (MQW) structures, it is referred to as the quantum Stark effect. Both are collectively called the electro-absorption effect, and optical modulators using this effect are known as electro-absorption modulators (EAMs).

Figure 2.21 qualitatively explains the electro-absorption effect. By inserting an optical modulation layer between a p-type and an n-type semiconductor and designing the impurity concentration of the optical modulation layer to be sufficiently low, a reverse voltage applied to the PN junction will concentrate the electric field in the depleted optical modulation layer. When light with a wavelength longer than the absorption edge wavelength (i.e., energy $h\nu$ smaller than E_g) is incident on the semiconductor without an applied external voltage, the light passes through the semiconductor without interacting with electrons in the valence band (ON). However, when a reverse voltage is applied, the bandgap tilts as shown in Fig. 2.21, causing the wavefunctions of the electrons to extend into the bandgap, reducing the energy difference between the conduction band and valence band to be equivalent to less than the bandgap energy E_g. When light with a wavelength longer than λ_g and energy smaller than E_g is incident, electrons transition from the valence band to the conduction band, resulting in light absorption (OFF). In EAMs, the intensity of the output light is directly modulated by controlling the applied voltage. The carriers (holes and electrons) generated in the valence and conduction bands due to optical absorption are quickly swept out to the external circuit as a photocurrent due to the electric field applied to the crystal, allowing for high-speed operation.

This electro-absorption effect has a high rate of change among phenomena where optical constants vary with external electrical signals, enabling intensity modulation with low voltage even with relatively short device lengths of around 200 μm.

(1) EA Modulators in the 1.5 μm Wavelength Band

Optical modulators with minimal dynamic spectral broadening are especially crucial for long-distance or high-speed optical transmission systems operating in the 1.5 μm

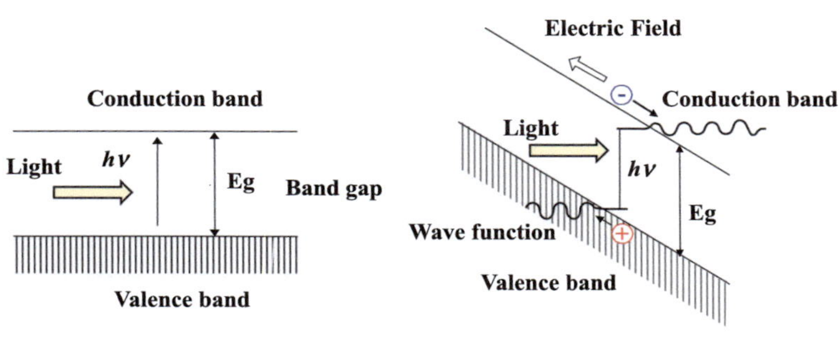

Fig. 2.21 Electro-absorption effect in semiconductor

wavelength band. The materials used for EAMs in this wavelength range are primarily InGaAsP material system, similar to semiconductor lasers.

When designing EAMs, three fundamental considerations should be addressed: reduction of operating voltage, reduction of insertion loss, and broadening of the bandwidth. From the perspective of low-voltage operation, it is advantageous to minimize the difference $\Delta E_g = E_g - hv$ between the optical energy hv and the bandgap energy E_g. However, this approach increases the absorption loss due to the built-in electric field when no voltage is applied. Furthermore, EAMs can be considered as capacitors electrically, and when terminated with 50 Ω for high-speed operation, they experience bandwidth limitation due to the CR time constant. Therefore, to achieve broad bandwidth, it is necessary to shorten the device length and increase the thickness of the waveguide, which contrasts with the requirements for low-voltage operation. This requires careful waveguide design to balance these competing requirements.

In addition to these basic design considerations, EAMs must also account for dynamic spectral broadening similar to direct-modulation lasers. When modulating optical intensity by changing the absorption coefficient, the refractive index also changes according to the Kramers–Kronig dispersion relations, leading to additional phase modulation. The spectral broadening of an EAM can be evaluated using the linewidth broadening factor α of semiconductor lasers.

α is defined as:

$$\alpha = \mathrm{d}n_r/\mathrm{d}n_i \tag{2.3.2}$$

which is the change in the ratio of the real and imaginary part of the complex refractive index $n_r + jn_i$. For Gaussian waveform pulses, the spectral broadening is approximately $\sqrt{1+\alpha^2}$. For 1.5 μm band semiconductor lasers, the α value is typically reported to be between 6 and 7. However, in EA modulators, by minimizing ΔE_g and increasing the electric field strength, it is possible to reduce the α value to below 1. Under these constraints, ΔE_g is usually designed to be between 30 and 50 meV, the waveguide layer thickness between 0.2 and 0.4 μm, and the device length between 100 and 200 μm.

Figure 2.22 shows an example of the structure and voltage characteristics of an EAM. In this example, the optical modulation layer uses InGaAsP (absorption edge wavelength $\lambda_g = 1475$ nm), with an energy difference ΔE_g between the bandgap energy and the optical energy at 1.55 μm being about 40 meV. The typical thickness, width, and length of the waveguide layer are approximately 0.3, 2, and 200 μm, respectively. With a voltage of − 2 V, an extinction ratio of over 20 dB and a modulation bandwidth exceeding 20 GHz have been achieved. In MQW structure EAMs, bandwidths exceeding 40 GHz have also been realized.

(2) Integrated Light Sources Combining Laser Diodes and Optical Modulators

InGaAsP-based EAMs are made from the same material system as 1.5 μm wavelength InGaAsP semiconductor lasers, allowing for monolithic integration with semiconductor lasers on a common InP substrate [4, 5]. Figure 2.23 shows a photograph of an

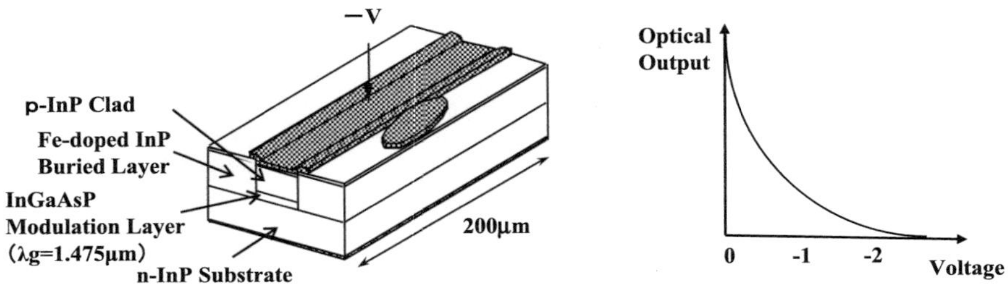

Fig. 2.22 Structure and typical characteristics of InGaAsP EAM

integrated light source. The DFB laser section consists of an InGaAsP $\lambda/4$ shift DFB laser, which is excellent for single-mode operation, while the EAM is also composed of InGaAsP. The lengths of the laser and modulator sections are 300 and 200 μm, respectively, achieving a compact laser-modulator integrated light source with the excellent optical coupling efficiency between the laser and the modulator.

EAM integrated lasers are suitable for miniaturization and cost reduction, making them ideal for long-distance optical submarine cable systems, such as the Japan Information Highway (JIH) cable, China-US cable, and other 2.5 Gb/s wavelength division multiplexed optical submarine cable systems. More recently, they have been widely used in transmitters for Fiber to the Home (FTTH) systems operating at 10 Gb/s and for data center systems. Additionally, standalone EAMs, which offer better wavelength stability compared to integrated light sources, are employed in long-distance optical submarine cables using 10 Gb/s wavelength division multiplexing, such as the Japan-US Trans-Pacific optical submarine cable and the Trans-Atlantic Telecommunication Cable-14 (TAT-14) across the Atlantic Ocean.

Fig. 2.23 Integrated light sources with $\lambda/4$ shift DFB laser and EA modulators

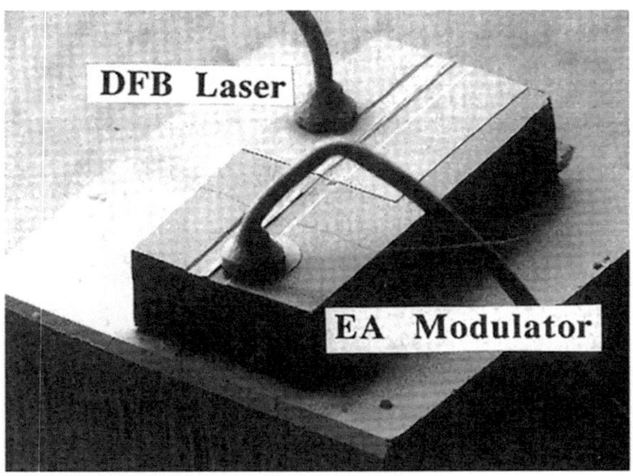

2.4 Photodetector

Photodetectors are used in receivers to convert optical signals into electrical signals. In optical-fiber communication, PIN photodiodes (PIN-PDs) or avalanche photodiodes (APDs) are commonly used. Photodetectors utilize the phenomenon where light is absorbed, generating photocurrent when light with a wavelength shorter than the bandgap energy of the semiconductor is incident. For materials, Si is used in the 0.8 μm band, while Ge and InGaAs are used in the 1.3–1.55 μm band.

As shown in Fig. 2.24, a PIN-PD has a structure in which an I-type semiconductor layer (Intrinsic Layer) is sandwiched between a PN junction. The p-region is made thin to reduce light absorption in this region, while the i-region is made thick so that most of the light is absorbed in this area.

When a reverse bias voltage with the negative voltage on the p-side and the positive voltage on the n-side of the PIN-PD is applied as shown in Fig. 2.25, light incident on the device is absorbed mainly in the i-region. This absorption generates electron-hole pairs proportional to the intensity of the incident light. The electron-hole pairs are accelerated by the electric field in the i-region, with electrons flowing toward the n-side and holes toward the p-side, resulting in a current.

On the other hand, the avalanche photodiode (APD) has a basic structure similar to that of the PIN photodiode but possesses a function that amplifies weak detected photocurrents by applying a large reverse bias voltage as shown in Fig. 2.26. This is

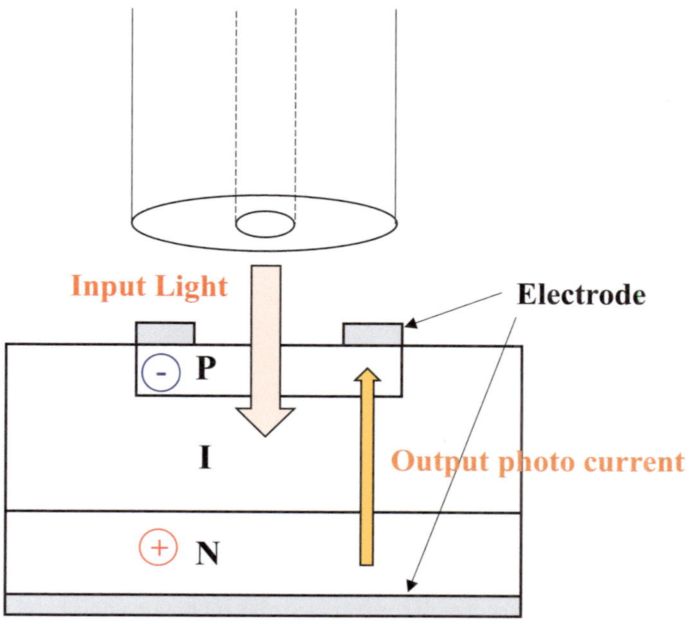

Fig. 2.24 Structure of PIN photodiode

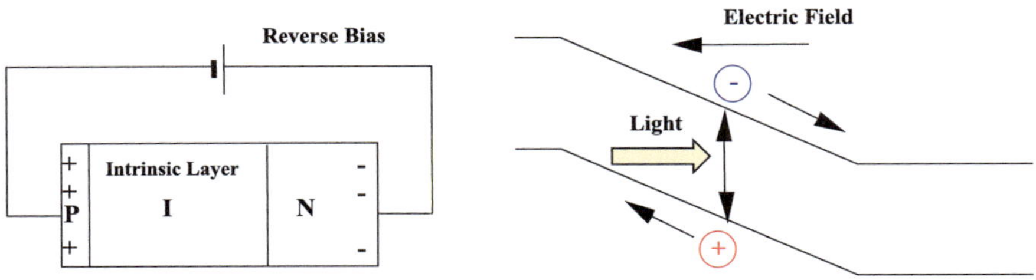

Fig. 2.25 Operation principle of PIN photodiode

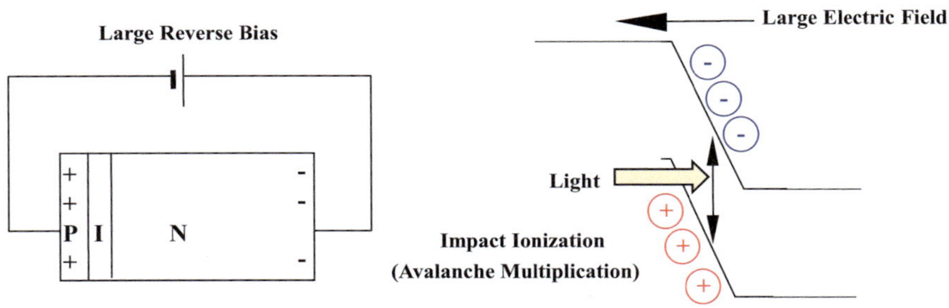

Fig. 2.26 Operation principle of avalanche photodiode

due to the avalanche multiplication effect, where carriers accelerated by the increased electric field collide with the lattice, generating electron-hole pairs. These generated electrons and holes then further collide with the lattice, creating additional electron-hole pairs. However, since APDs generate excess noise, the receiver must be designed to operate at optimal gain when using APDs.

In trans-oceanic submarine cable systems employing regeneration repeaters, such as TPC-3 and TPC-4, APDs were used as photodetectors in repeaters. On the other hand, in optical submarine cable systems using optical amplifier repeaters, such as TPC-5 and later, PIN-PDs were used in receivers. This is becaluse, in such systems, noise accumulates along the transmission path, making noise reduction essential. The system performance is determined not by receiver sensitivity but by the SNR caused by amplifier noise accumulation in the transmission path, and therefore, PIN-PDs, which do not generate excess noise like the APD, are used in receivers. Currently, APDs are used to extend transmission distances in short-haul systems without optical amplifiers, such as Passive Optical Networks (PON) and Ethernet transceivers.

In intensity modulation schemes where "0" and "1" are transmitted based on differences in light intensity, the electrical signal converted by the photodetector can be restored to "0" and "1" codes by using a threshold to determine the voltage. When the received optical signal is free from noise and waveform distortion, it can be received without errors. However, if the received signal contains noise and waveform distortion, errors may occur with a certain probability. This probability is referred to as the bit error rate (BER). When the received signal contains noise, the histogram

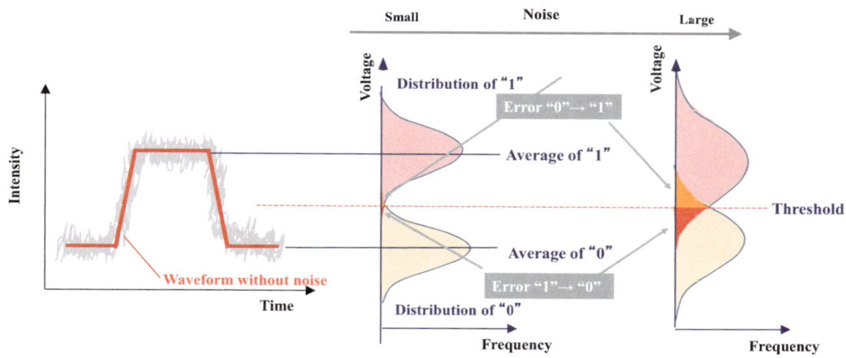

Fig. 2.27 Decision for received signals

of the voltage becomes like the right side of Fig. 2.27. Due to noise, there is a probability that a signal meant to be "1" ("0") may have a voltage smaller (larger) than the threshold, leading to a bit error. The number of bit errors increases with the level of noise.

One commonly used indicator for defining the characteristics of a receiver is the minimum optical sensitivity. This is determined by varying the input optical signal power to the receiver using a variable optical attenuator while measuring the BERs using pseudo-random binary sequences (PRBS). The minimum optical sensitivity is indicated by the receiver power at which a BER of either 1×10^{-9} or 1×10^{-12} is achieved. Figure 2.28 shows examples of minimum optical sensitivity measurements for 400 Mb/s and 1.6 Gb/s signals. In this example, the minimum optical sensitivity for a 400 Mb/s signal with a BER of 1×10^{-9} is − 37 dBm, while for a 1.6 Gb/s signal, it is − 32.4 dBm. Here, the unit [dBm] used for received optical power is expressed relative to 1 mW, and the signal power of X [mW] is converted to [dBm] using $10 \log_{10}(X)$. When increasing the bit rate from 400 Mb/s to 1.6 Gb/s, the signal bandwidth is increased by a factor of 4, which also increases the noise by a factor of 4. Therefore, to achieve the same BER, the received signal power needs to be increased by a factor of 4 (6 dB) to maintain the same optical signal-to-noise ratio (optical SNR). In the example shown in Fig. 2.28, the difference in minimum optical sensitivity is less than 6 dB, indicating that the receiver for the 1.6 Gb/s signal needs further optimization.

In optical communication systems, the Q value is also commonly used as an indicator of signal quality, in addition to the BER. This is because, especially for optical signals with lower bit rate, measuring a BER such as 1×10^{-12} requires a long time, and using the Q value, which is expressed in decibels, makes quantitative comparison of characteristics easier. The Q value represents the SNR assuming that the probability density function of the received signal is Gaussian. As shown in Fig. 2.29a, if the mean voltages of the signals "1" and "0" after conversion to electrical signals are μ_1 and μ_0, and the standard deviations of the voltage variations due to

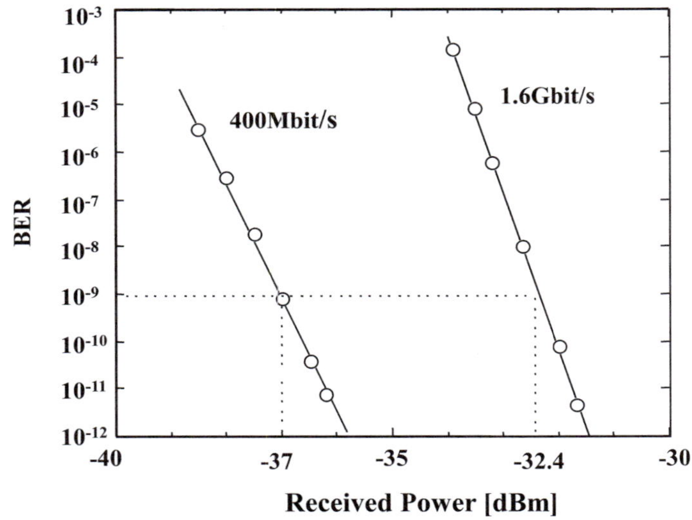

Fig. 2.28 Example of receiver sensitivity measurement

noise are σ_1 and σ_0, respectively, the Q value is expressed by the following formula.

$$Q \equiv \frac{|\mu_1 - \mu_0|}{\sigma_1 + \sigma_0} \tag{2.4.1}$$

The Q value is often expressed in decibels, calculated as $10 \log(Q^2)$. For example, when optical SNR is the dominant factor determining signal quality, an increase of 1 dB in the optical SNR results in an increase of 1 dB in the Q value.

To measure the Q value, one method involves using a digital sampling oscilloscope to measure the received signal waveform, then determining μ_1, μ_0, σ_1, and σ_0 from the measured waveform and calculating the Q value using Eq. 2.4.1. However, this method can lead to inaccuracies in σ_1 and σ_0 due to the signal pattern effects and

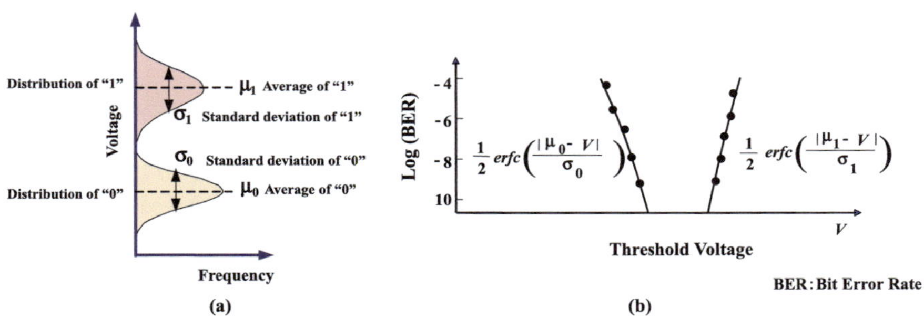

Fig. 2.29 Measurement method for Q-factor

oscilloscope noise impacting the measurement results. Therefore, a more commonly used method [6], as shown in Fig. 2.29b, involves measuring the BER by varying the decision threshold and deriving the Q value from the approximate straight lines of the "1" and "0" sides.

BER and Q-value are converted using the following approximate formula:

$$\mathrm{BER} = \frac{1}{2}\,\mathrm{erfc}\,\frac{Q}{\sqrt{2}} \approx \frac{\exp\left(-Q^2/2\right)}{Q\sqrt{2\pi}} \tag{2.4.2}$$

In recent transmission systems, the advanced Forward Error Correction schemes explained in the next section are commonly used. In these cases, the BERs before the error correction are relatively large, thus can be directly measured without the method shown in Fig. 2.29b.

2.5 Forward Error Correction

In a communication system, when the received signal's SNR is insufficient, bit errors occur. As shown in Fig. 2.30, the method of adding redundant information to the data signal at the transmission side and using that redundant information to correct errors at the reception side is known as Forward Error Correction (FEC).

The first application of FEC in optical communication systems was in the trans-oceanic optical submarine cable systems of the 1990s. When the optical amplifier

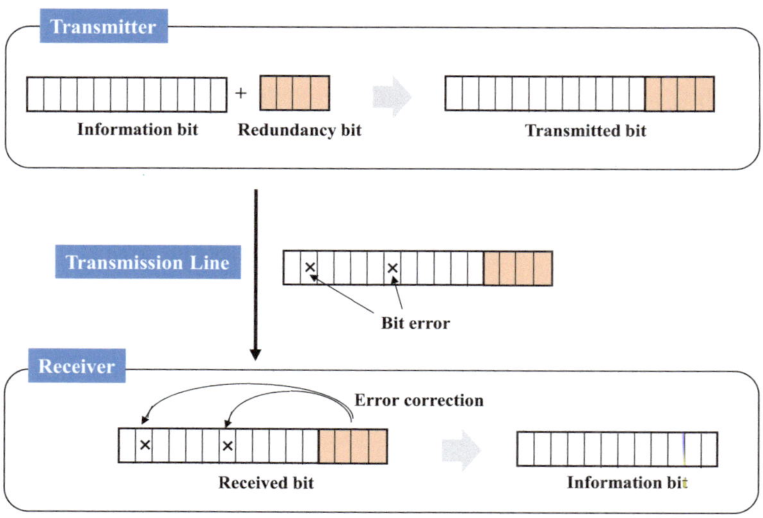

Fig. 2.30 Forward error correction

repeater was introduced in TPC-5, it was discovered that transmission characteristics varied due to polarization fluctuations. As a solution, the Reed–Solomon code RS(255,239), well-known in wireless communication systems, was implemented and confirmed to effectively suppress transmission characteristic variations [7]. RS(255,239) codes, after being applied to trans-oceanic optical submarine cable systems, became widely used in long-haul optical communication systems and were standardized internationally as ITU-T Recommendation G.975. Subsequently, it has become an essential technology for ensuring system margin, especially in long-haul backbone optical transmission systems.

The RS(255,239) code is referred to as the first-generation FEC applied to optical communication and consists of adding 16 redundant bits to 239 information bits. Thus, the redundancy rate is $16 \div 239 \approx 7\%$. By using this code, it is possible to improve a BER of 1.4×10^{-4} to 1×10^{-13}. When converting BER to Q value, BERs of 1.4×10^{-4} and 1×10^{-13} corresponds to Q values of 11.2 and 17.3 dB, respectively. Therefore, the coding gain of this code is 6.1 dB. After subtracting the increase due to redundant bits $10 \log(1.07) = 0.3$ dB, the net coding gain is 5.8 dB. The greater the redundancy of FEC, the higher its error correction capability. However, since the required bandwidth also increases accordingly, the net coding gain is an important performance metric for FEC.

For the second generation of FEC, concatenated codes and iterative decoding were introduced to improve error correction capabilities. Figure 2.31 illustrates the structure of concatenated codes and iterative decoding methods. In the transmitter, the information bit encoded with the first code (outer code) is further encoded with the second code (inner code). At the receiver, errors that could not be corrected in a single iteration are addressed by repeatedly decoding from the inner code to the outer code. Various combinations of codes have been proposed as second-generation FEC, achieving net coding gains of around 8–9 dB. Second-generation FEC became widely used in wavelength division multiplexing technologies adopted in the late 1990s, dealing with unavoidable degradation in SNR due to increased wavelength channels.

In the third generation of FEC, further improvement in coding gain was achieved by introducing soft decision decoding. While hard decision decoding uses a single threshold to determine a 1 or 0, soft decision decoding uses multiple thresholds to provide confidence information indicating the likelihood of a 1 (or 0), and then performs error correction. Turbo codes and Low-Density Parity-Check (LDPC) codes are examples of codes that provide high error correction capability through soft decision decoding. Turbo codes with concatenated BCH codes, such as BCH(144,128) $\times$ BCH(256,239), have demonstrated a net coding gain of 10.1 dB with a redundancy of 23.9% for a 12.4 Gbit/s signal [8]. LDPC codes, which approach the Shannon limit for high error correction capabilities and are suitable for parallel implementation, have been developed for FEC in 40 and 100 Gb/s transmission systems. By using an inner code with soft decision LDPC codes (redundancy: 13%) and an outer code with hard decision block codes (redundancy: 7%), a concatenated code with an overall redundancy of 20.5% and performing 16 iterations of decoding, net coding gains exceeding 10 dB have been achieved [9]. Figure 2.32 is comparing the error

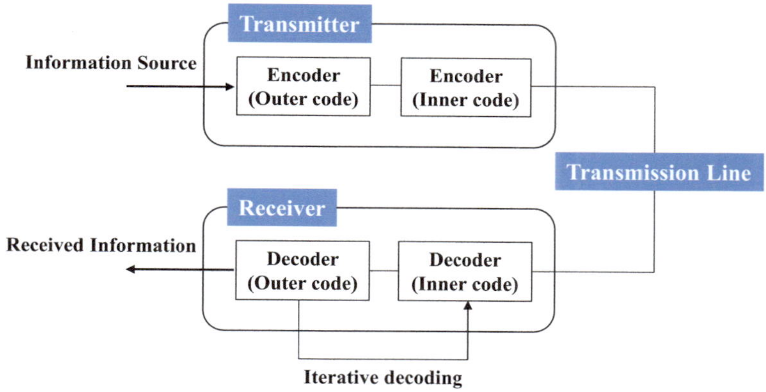

Fig. 2.31 Configuration of concatenated code with iterative decoding

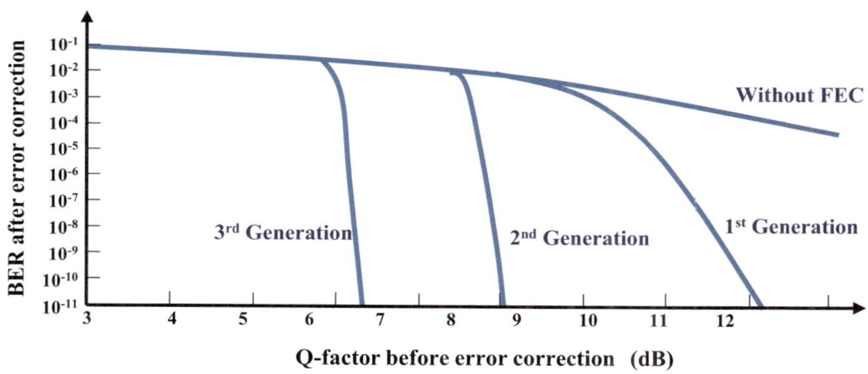

Fig. 2.32 Characteristic example of FEC

correction performance of the first to third generation FEC used in optical fiber communication systems.

2.6 Optical Repeater

2.6.1 Types of Optical Repeaters

Repeaters are used to restore attenuated optical signals in optical fiber transmission and relay them to the next section of fiber. They can be broadly classified into regeneration repeaters and optical amplifier repeaters. Figure 2.33a illustrates the functional block diagram of a regeneration repeater. In a regeneration repeater, the

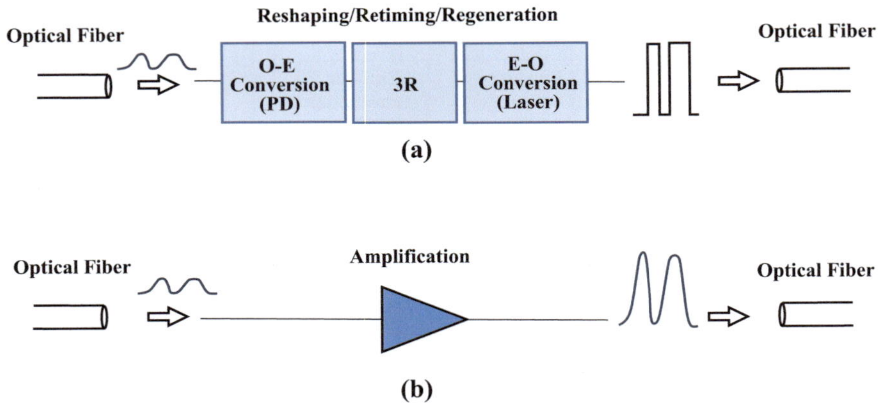

Fig. 2.33 Optical repeaters

optical signal is first converted to an electrical signal using a photodetector (O–E conversion). Next, waveform equalization (Re-shaping) is performed using a low-pass filter, timing synchronization (Re-timing) is achieved through clock extraction, and Re-generation is done after signal decision. These processes, "3R functions," are carried out in the electrical domain. Subsequently, the electrical signal is converted back to optical form using a laser (E–O conversion). In regeneration repeaters, since digital signals are regenerated in the electrical stage, all types of waveform degradation are reset, and the optical signal is fully regenerated. Note that the bit error after signal decision propagates to next section. Therefore, system design only needs to consider the regeneration repeater sections, which typically range from about 50 to 150 km. This method was implemented in the first Trans-Pacific optical submarine cable systems, TPC-3 and TPC-4, as discussed in Chap. 4.

Figure 2.33b shows an optical amplifier repeater system. In this system, the attenuated optical signal in optical fiber transmission is directly amplified by the optical amplifier. As a result, amplifier noise and various types of optical waveform degradation due to chromatic dispersion and polarization dispersion accumulate in the transmission. Therefore, when using optical amplifiers, it is necessary to design the entire system to account for these accumulated degradations.

2.6.2 *Types of Optical Fiber Amplifiers*

Since the introduction of the Erbium-Doped Fiber Amplifier (EDFA) in the 1980s, practical research and development of optical fiber amplifiers have been conducted worldwide. Optical fiber amplifiers are categorized based on the type of dopant used. Figure 2.34 shows the amplification bands of various optical fiber amplifiers made

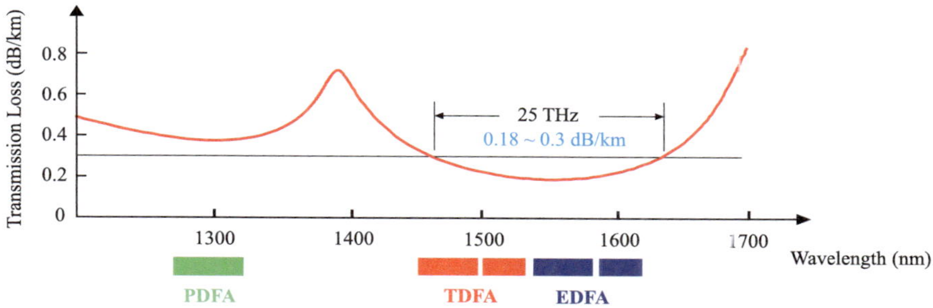

Fig. 2.34 Fiber attenuation characteristic and amplification bands of optical fiber amplifiers

from silica fibers. Table 2.4 lists the names of each wavelength band. Praseodymium-Doped Fiber Amplifiers (PDFA) operate in the 1300 nm band (O band). Thulium-Doped Fiber Amplifiers (TDFA) operate in the 1400 nm band (S band). Erbium-Doped Fiber Amplifiers (EDFA) operate in the 1550 nm band (C band) and the 1600 nm band (L band). Additionally, fluoride fiber-based amplifiers have also been commercialized and are mainly used for amplification in the 1300 nm band.

(1) EDFA

In long-distance optical fiber communication systems, EDFA has been used in all systems since TPC-5, which is discussed in Chap. 5. As shown in Fig. 2.35, when light at wavelengths of 1480 nm or 980 nm is input into an Erbium-Doped Fiber (EDF), the input optical signal at 1550 nm (C band) is amplified.

A detailed explanation is provided in Chap. 5, but optical amplification in EDFA occurs as illustrated in Fig. 2.36. Electrons at the ground state $N1$ are excited to a higher energy level by 980 nm or 1480 nm light. These electrons then transition to the energy level $N2$ (where the energy difference between $N1$ and $N2$ corresponds to the wavelength of 1550 nm) and are subsequently stimulated to return to the ground state $N1$, emitting light and resulting in amplification. Spontaneous emission light that does not contribute to amplification becomes optical noise (Amplified Spontaneous

Table 2.4 Wavelength band

Band	Description	Range (nm)
O-band	Original	1260 ~ 1360
E-band	Extended	1360 ~ 1460
S-band	Short wavelength	1460 ~ 1530
C-band	Conventional	1530 ~ 1565
L-band	Long wavelength	1565 ~ 1625
U-band	Ultra long wavelength	1625 ~ 1675

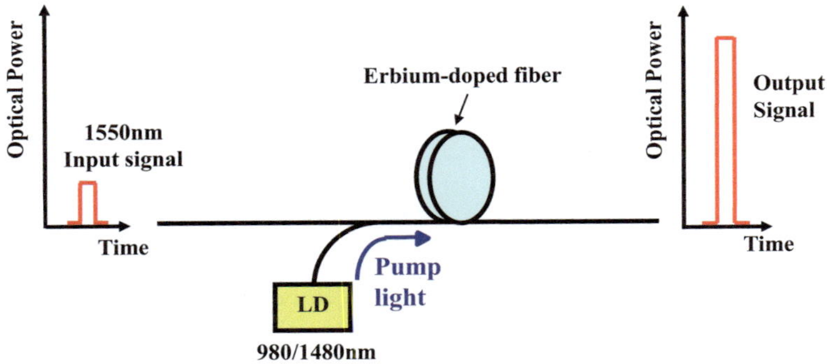

Fig. 2.35 Basic configuration of EDFA and signal amplification

Emission noise: ASE noise). The ratio of electrons excited from $N1$ to $N2$ is known as the population inversion ratio (the reciprocal of the amplified population inversion coefficient n_{sp}).

Figure 2.37 shows the wavelength dependence of the gain coefficient per unit length of EDFA with the average population inversion ratio as a parameter. With 980 nm pumping, all electrons in $N1$ can be excited to $N2$, achieving a 100% average population inversion ratio and resulting in strong gain peaks near 1536 nm. This results in minimal ASE noise because $n_{sp} = 1$. In contrast, with 1480 nm pumping, where complete inversion cannot be achieved (for example, 80% population inversion ratio as shown in the figure), a relatively smoother amplification characteristic is obtained. In the first optical submarine cable system with optical amplifiers (TPC-5, see Chap. 5), 1480 nm pumped C band EDFA was used. Since the 10 Gbit/s WDM system (see Chap. 7), C band EDFA with 980 nm pumping, which has superior noise characteristics, has been used.

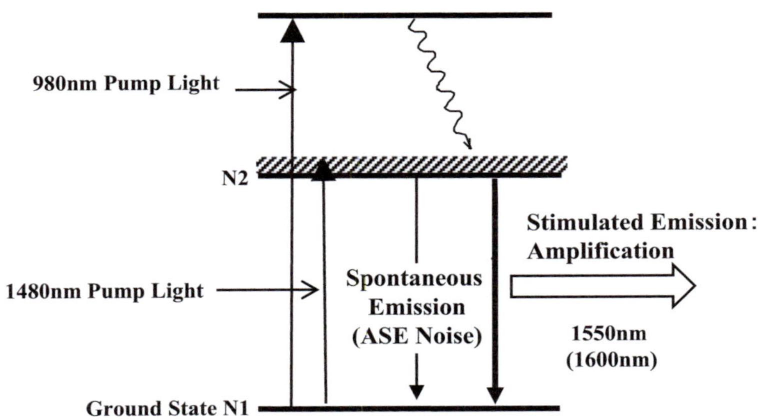

Fig. 2.36 Optical amplification process in EDFA

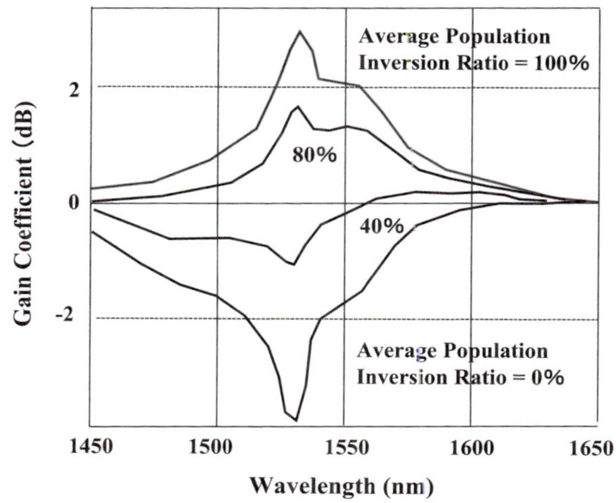

Fig. 2.37 Wavelength dependence of gain coefficient

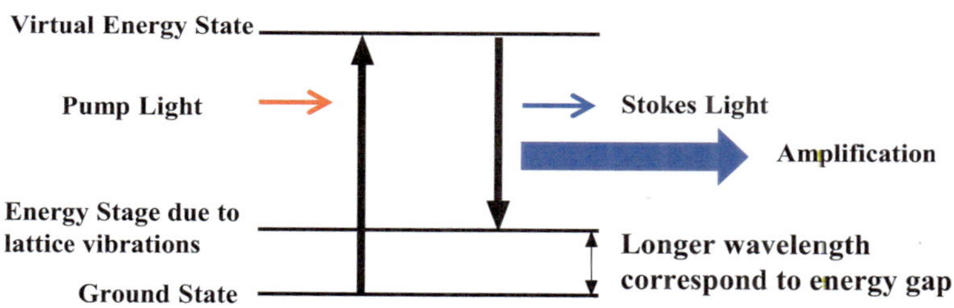

Fig. 2.38 Raman amplification process

Conversely, when the population inversion ratio is around 40%, as shown in Fig. 2.37, a small gain peak shifts to near 1600 nm. With longer EDF to increase the gain, an L-band amplifier can be realized. L-band amplifiers are mainly used in terrestrial optical communication systems to shift the amplification band to longer wavelengths, achieving limited chromatic dispersion for WDM transmission in DSF.

(2) Raman Amplifier

While EDFA is a lumped-type optical amplifier, another representative optical fiber amplifier is the distributed Raman fiber amplifier. As shown in Fig. 2.38, when strong pump light is introduced into the silica glass, interactions between the silica crystal lattice vibrations and the pump light generate scattered light (Stokes light) about 100 nm longer in wavelength, resulting in stimulated Raman Scattering. When an optical signal at the same wavelength band as the Stokes light is launched in the optical fiber, it is amplified by this stimulated Raman Scattering. As described in Sect. 2.1, pure silica core fibers have the gain peaks at a frequency approximately

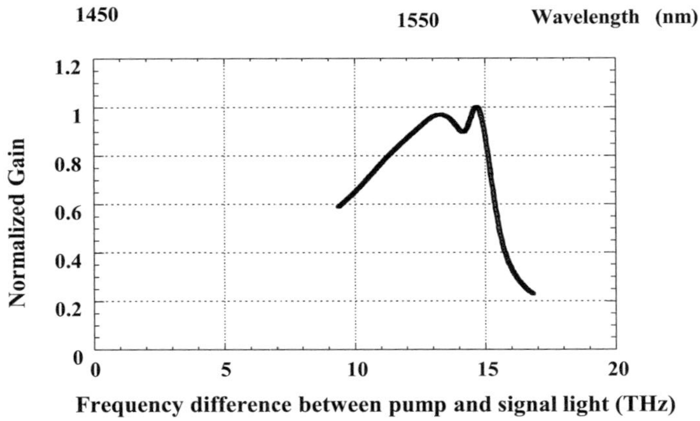

Fig. 2.39 Example of Raman gain in pure silica-core fiber

1.32 THz lower than the pump light frequency (about 100 nm longer wavelength in the 1550 nm band). Figure 2.39 shows a measurement example of normalized Raman gain in a pure silica core fiber with a 1450 nm pump light. It is evident that the optical amplification gain peaks at the 1550 nm wavelength band, approximately 100 nm away from the 1450 nm pump light.

In Raman amplification, the optical fiber itself serves as the optical amplifier, allowing for an effectively lossless optical fiber transmission line as shown in Fig. 2.40. The efficiency of Raman amplification increases with lower fiber losses and larger fiber nonlinearity. One of the key features of Raman amplifiers is the ability to freely set the amplification band based on the wavelength of the pump light. It is also possible to expand the wavelength range by using a hybrid configuration of EDFA and Raman amplification. Additionally, Raman amplifiers have a low equivalent noise figure, which allows for a reduction in the input power of the optical signal into the fiber, thereby mitigating the fiber nonlinearity and making them suitable for high-speed signal transmission.

Problems

2.1 Calculate the output signal power from 20 km optical fiber with the attenuation of 0.3 dB/km when the fiber-launched power is 10 mW.

2.2 Calculate the output signal power when the fiber attenuation is reduced to 0.15 dB/km from 2.1.

2.3 Calculate the arriving time difference when two optical pulses with the wavelength difference of 1 nm are simultaneously launched to 10 km optical fiber with chromatic dispersion of 20 ps/nm/km.

2.4 Calculate the bit error rate corresponding to Q factor of 15.6 dB.

2.5 Calculate the FEC gain for a FEC code which can improve the bit error rate from 1.4×10^{-4} to 8.6×10^{-10}.

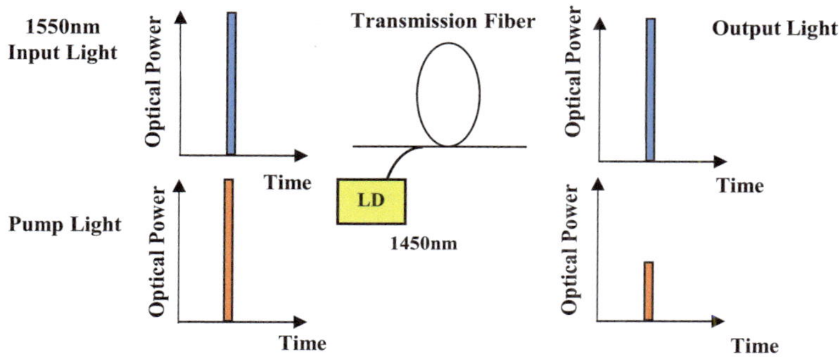

Fig. 2.40 Loss-less fiber transmission by Raman amplifier

References

1. S. Sato, Y. Kawaguchi, H. Sakuma, T. Haruna, T. Hasegawa, Record low loss optical fiber with 0.1397 dB/km, in *OFC2024, Tu2E.1* (2024)
2. G.P. Agrawal, *Nonlinear Fiber Optics* (Academic Press)
3. S. Akiba, M. Usami, K. Utaka, 1.5-μm λ/4-shifted InGaAsP/InP DFB lasers. IEEE J. Lightwave Technol. **5**(11), 1564–1573 (1987)
4. M. Suzuki, H. Tanaka, H. Taga, S. Yamamoto, Y. Matsushima, λ/4-shifted DFB laser/electroabsorption modulator integrated light source for multigigabit transmission. IEEE J. Lightwave Technol. **10**, 90–95 (1992)
5. M. Suzuki, Y. Noda, H. Tanaka, S. Akiba, Y. Kushiro, H. Isshiki, Monolithic integration of InGaAsP/InP distributed feedback laser and electroabsorption modulator by vapor phase epitaxy. IEEE J. Lightwave Technol. **5**, 1277–1285 (1987)
6. N. Bergano, F. Kerfoot, C. Davidson, 5 Gbit/s optical transmission terminal equipment using forward error correcting code and optical amplifier. IEEE Photon. Technol. Lett. **5**(3), 304–306 (1993)
7. S. Yamamoto, H. Takahira, M. Tanaka, 5 Gbit/s optical transmission terminal equipment using forward error correcting code and optical amplifier. Electron. Lett. **30**(3), 254–255 (1994)
8. T. Mizuochi, Recent progress in forward error correction and its interplay with transmission impairments. IEEE J. Sel. Top. Quantum Electron. Opt. Commun. **12**(4), 544–554 (2006)
9. T. Mizuochi, T. Sugihara, Y. Miyata, K. Kubo, K. Onohara, S. Hirano, H. Yoshida, T. Yoshida, T. Ichikawa, Evolution and status of forward error correction, in *OFC/NFOEC2012, OTu2A.6* (2012)

Chapter 3
Modulation Formats

Abstract In optical communication systems, intensity modulation with direct detection (IM-DD) scheme has been used for a long time. However, as it became challenging to achieve high-speed transmission with this method alone, phase modulation was introduced. Early phase-modulated transmission systems used differential encoding and delay interferometers to convert phase-modulated signals into intensity-modulated signals for detection, but this approach imposed limitations on the types of modulation formats that could be used. To overcome these limitations and introduce more advanced modulation formats such as quadrature amplitude modulation (QAM) for increasing a spectral efficiency, the adoption of coherent detection became necessary. This chapter explains these modulation formats employed in long-haul transmission systems.

Keywords Intensity modulation · Direct detection · On-Off Keying · Non-Return to Zero pulse · Return to Zero pulse · Pulse Amplitude Modulation · Differential Phase Shift Keying · Quadrature amplitude modulation

3.1 Intensity Modulation

In optical communication systems, Intensity Modulation-Direct Detection (IM-DD) has been used for a long time due to its simplicity. Intensity modulation encodes information onto the intensity of light by using direct modulation of a laser or an external modulator as explained in Chap. 2. Binary intensity modulation is also referred to as On-Off Keying (OOK). Since the photo detectors can detect the intensity of light, the receiver for OOK signals simply consists of a photo detector. OOK signal can be implemented using two formats: NRZ (Non-Return to Zero)-OOK and RZ (Return to Zero)-OOK. The temporal waveforms of NRZ-OOK and RZ-OOK signals are shown in Fig. 3.1. In NRZ-OOK, rectangular pulses with a width equal to the bit interval are used, whereas in RZ-OOK, the pulse width is approximately half of the bit interval. In NRZ-OOK signals, if the bit "1" is repeated n times, the pulse width extends n times; in RZ-OOK signals, repeated "1"s appear as isolated pulses

M. Suzuki and I. Morita, *Long-Haul Optical Fiber Communication Systems*, Textbooks in Telecommunication Engineering, https://doi.org/10.1007/978-3-032-19472-5_3

consecutively. For RZ-OOK signals, additional optical modulator is required for RZ pulse generation. Therefore, the transmitter configuration of RZ-OOK signal is more complicated compared with NRZ-OOK signals, however, tolerance to the fiber nonlinearity is larger in RZ-OOK signals as explained in Chap. 6.

Figure 3.2 shows constellation diagram of OOK signal, which shows the real and imaginary parts of the optical electric fields of the signal. In OOK signals, since the phase of the light is not used for data modulation, the signal point for data "1" can be plotted anywhere on the circle in the constellation diagram.

To increase the transmitted data with intensity modulated signals without increasing the symbol rate, it is effective to use Pulse Amplitude Modulation (PAM), which employs multi-level pulse amplitudes as shown in Fig. 3.3. PAM4 and PAM8 are modulation formats that use 4 and 8 amplitude levels, respectively. PAM4 can transmit 2 bits per symbol, while PAM8 can transmit 3 bits per symbol. These signals are employed in recent short-haul communication systems, such as 400 Gb/s Ethernet.

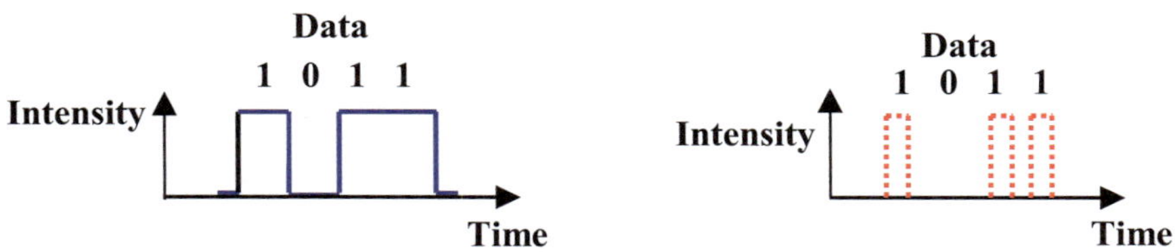

(a) NRZ signal (b) RZ signal

Fig. 3.1 Temporal waveforms of OOK signals

Fig. 3.2 Consternation diagrams of OOK signals

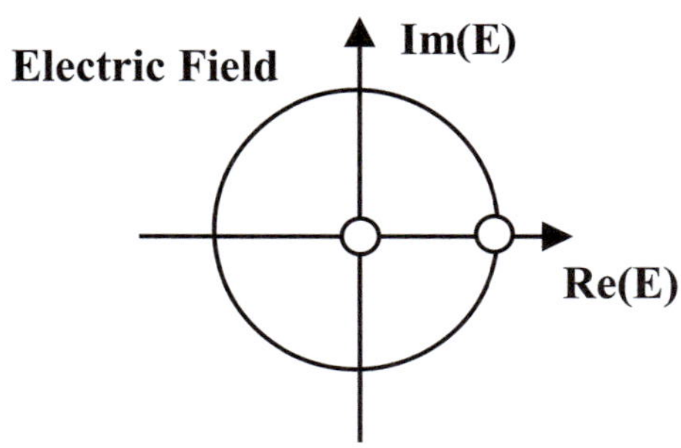

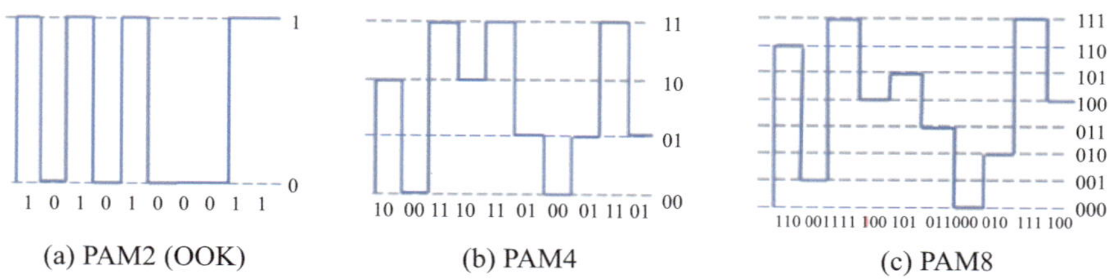

Fig. 3.3 Temporal waveforms of PAM signals

3.2 Phase Modulation

Phase modulation encodes information onto the phase of light. The method of detecting the phase of light by using interference with a local oscillator is called coherent detection and its basic configuration is shown in Fig. 3.4. The details of coherent detection will be discussed in Chap. 8.

In this chapter, we focus on Differential Phase Shift Keying (DPSK) and Differential Quadrature Phase Shift Keying (DQPSK), which encode phase information and then convert it to intensity modulated signals at the receiver. The converted signals can be detected with conventional photo detectors. Figure 3.5 shows the temporal waveforms of DPSK and DQPSK signals, respectively. In these figures, solid lines represent waveforms in the NRZ format, while dashed lines depict waveforms in the RZ format. In subsequent discussions, NRZ-formatted DPSK (DQPSK) will simply be referred to as DPSK (DQPSK), while RZ-formatted DPSK (DQPSK) will be denoted as RZ-DPSK (RZ-DQPSK). DPSK signals maintain a constant amplitude, with information encoded in the phase of 0 and π, carrying 1 bit per symbol. DQPSK signals also maintain constant amplitude but encode information in the phase of 0, $\pi/2$, π, and $3\pi/2$, allowing for 2 bits per symbol, making it a multi-level signal format. Figure 3.6 shows the constellation diagrams of DPSK and DQPSK signals.

DPSK modulates information in the phase of the light, which is then converted to an intensity modulated signal at the receiver side using a 1-bit delay interferometer.

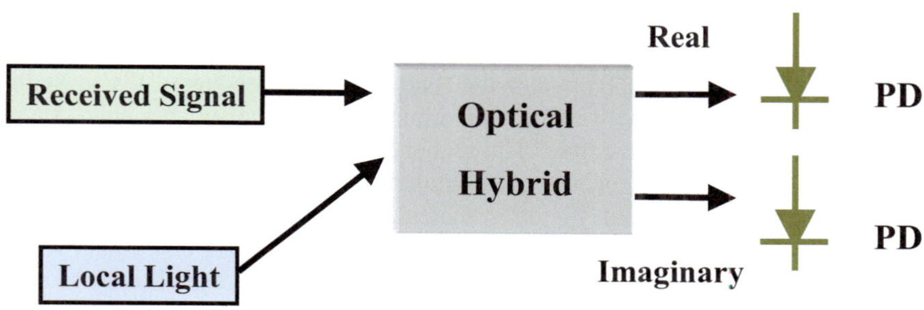

Fig. 3.4 Basic configuration of coherent receiver

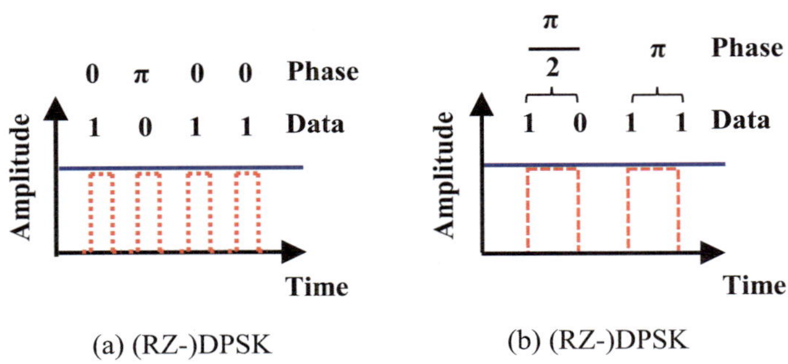

Fig. 3.5 Temporal waveforms of DPSK and DQPSK signals (solid line: NRZ signal, dashed line: RZ signal)

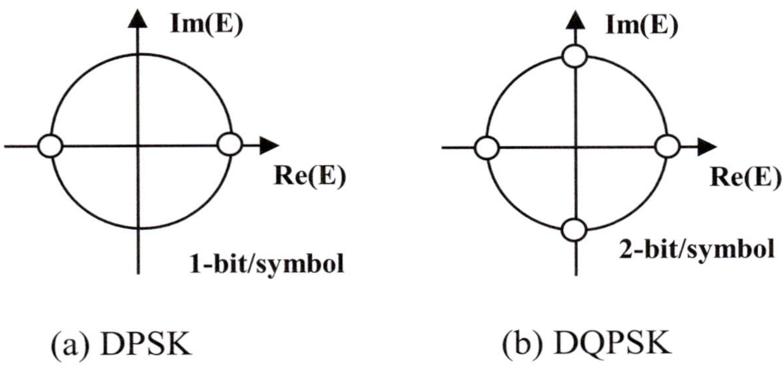

Fig. 3.6 Consternation diagrams of DPSK and DQPSK signals

To ensure the transmitted data matches the output of the delay interferometer at the receiver, pre-coding is performed at the transmitter. In the differential pre-coding process, intensity (1, 0) is converted into phase (π, 0) to generate phase modulation information. If a "1" (π) is received, the signal is inverted; if a "0" (0) is received, the state remains unchanged, resulting in a DPSK-modulated signal. Figure 3.7 shows an example of DPSK signal generation. In this example, starting with the first "1" (π) as a reference, the next "0" retains the state (π), and subsequent "1"s cause the state to alternate, generating 0 and π. This illustrates how the signal's phase states are altered based on the transmitted data sequence.

Figure 3.8 illustrates the receiver configuration for DPSK signals. When a DPSK signal is input into the 1-bit delay interferometer, two output ports provide the original data signal and its logical inversion. This effectively doubles the optical signal power due to the two-path interference, reducing the required optical signal-to-noise ratio (SNR) for the same bit error rate. Consequently, the receiver sensitivity improves by 3 dB compared to NRZ-OOK signals, allowing for lower transmitted optical power.

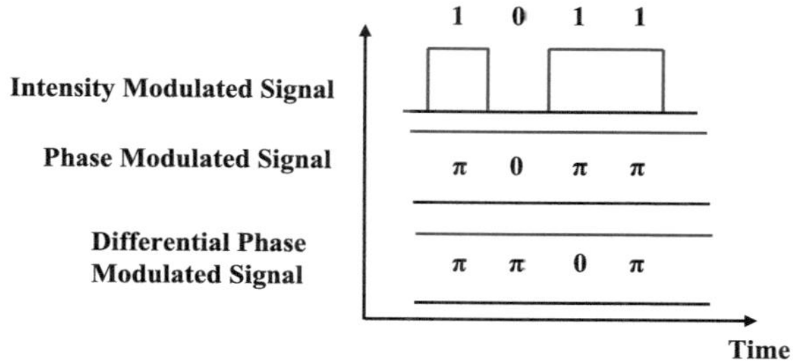

Fig. 3.7 Generation of DPSK signal

This reduction in transmitted power enhances the system's tolerance to optical fiber nonlinearities.

DQPSK modulation extends the principles of DPSK to include additional phase states: $\pi/2$ and $3\pi/2$, in addition to 0 and π. DQPSK carries 2 bits of information per symbol, effectively halving the symbol rate compared to DPSK with the same bit rate and thus reducing the spectral width to half of that of DPSK. This narrower spectral width enhances tolerance against chromatic dispersion and polarization mode dispersion (PMD). However, unlike DPSK, which requires two photo detectors, DQPSK necessitates four photo detectors, making the receiver configuration more complex compared to DPSK.

Another narrow-spectrum modulation format is Duo-Binary signaling, which generates intensity-phase-modulated signals through pre-coding. This signal can be received using a single photo detector to reconstruct the original intensity signal. The pre-coding for Duo-Binary signals involves converting binary input data (0, 1) into a ternary output series (0, 1, 2) using partial response coding that outputs (1, 1) for input 1. After removing the DC component, the series is further transformed into (− 1 (π), 0, 1). This format halves the bandwidth compared to NRZ-OOK signals, improving dispersion tolerance, but it has lower noise tolerance. Consequently, a higher optical SNR is required to achieve the desired bit error rate, and it is less resistant to fiber nonlinearities.

The key characteristics of various modulation formats which can be received without coherent detection are summarized in Table 3.1 [1]. This table includes temporal waveforms, spectral widths, modulator and receiver configurations, and overall features. For DPSK and DQPSK signals, the features of RZ-DPSK and

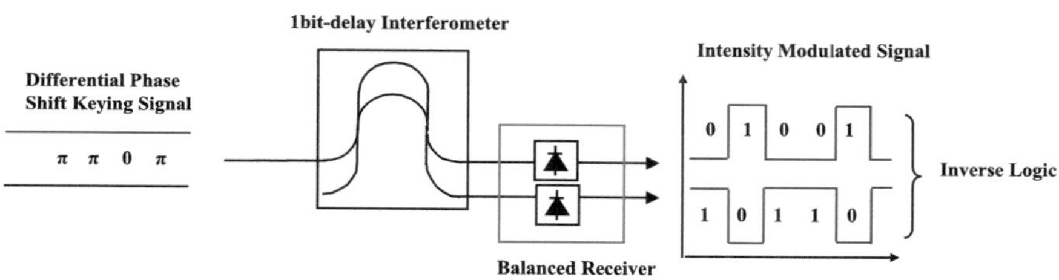

Fig. 3.8 Configuration of DPSK receiver

Table 3.1 Characteristics of various modulation formats

Modulation Format	Temporal Waveform	Frequency Spectrum	Optical Modulator	Optical Receiver	Features
Intensity Modulation (NRZ)	Optical Power; 1 0 1 1 0 0 1 1; Time	Frequency	Mach-Zehnder modulator; Data	PIN-PD	• Simplest configuration (Low cost) • Low tolerance to noise, CD and PMD
Intensity Modulation (RZ)	Optical Power; 1 0 1 1 0 0 1 1; Time	Frequency	Data; Clock (RZ)		• Simple configuration (Low cost) • Larger tolerance to fiber nonlinearity • Relatively large tolerance to noise/CD/PMD
Intensity Modulation (Duobinary)	Optical Power; 1 0 1 1 0 0 1 1; 0 π 0 Phase; Time	Frequency	Precoded Data (IM); LP		• Smaller signal bandwidth • Large tolerance to CD • Low tolerance to noise and fiber nonlinearity
RZ-DPSK	Optical Power; 1 0 1 1 0 0 1 1; 0 0 π 0 0 0 π 0 Phase; Time	Frequency	Precoded Data (PM); Clock (RZ)	Delay interferometer	• Larger tolerance to noise, PMD and fiber nonlinearity • Complex configuration
RZ-DQPSK	Optical Power; 1 0 1 1 0 0 1 1; π/2 π 0 π Phase; Time	Frequency	Precoded Data (PM); Clock (RZ); π/2; Precoded Data (PM)		• Halved symbol rate • Larger tolerance to noise, CD, PMD and fiber nonlinearity • Smaller signal bandwidth • More complex configuration

RZ-DQPSK are only shown in the table, since they exhibit resilience against fiber nonlinearity and are well-suited for long-distance optical transmission systems.

3.3 Quadrature Amplitude Modulation

Intensity and phase modulations encode information onto either the amplitude or the phase of light. On the other hand, quadrature amplitude modulation (QAM) encodes information both the amplitude and the phase of light. This requires coherent detection shown in Fig. 3.4. With QAM and coherent detection, multi-level modulation can be easily introduced. As an example of multi-level QAM signal, Fig. 3.9 shows the constellation diagrams of 64QAM, which uses 64 combinations of amplitude and phase. With multi-level modulation signals, which use multiple amplitude and phase states, more than one bit information can be transmitted with a single symbol. For example, 64QAM can transmit 6 bits of information with a symbol. With multi-level modulation, the symbol rate required to achieve the same bit rate can be reduced, and the signal bandwidth can also be reduced. This not only enables high-speed signal transmission and detection using low-speed optical and electrical components but also improves spectral efficiency in wavelength division multiplexing (WDM) systems.

To generate multi-level QAM signals, vector modulation is required to obtain any combination of amplitude and phase states. There are two main methods to realize vector modulation: one method is to independently modulate the intensity and phase of light using an intensity modulator and a phase modulator (as shown in Fig. 3.10a), and the other method uses two intensity modulators to independently

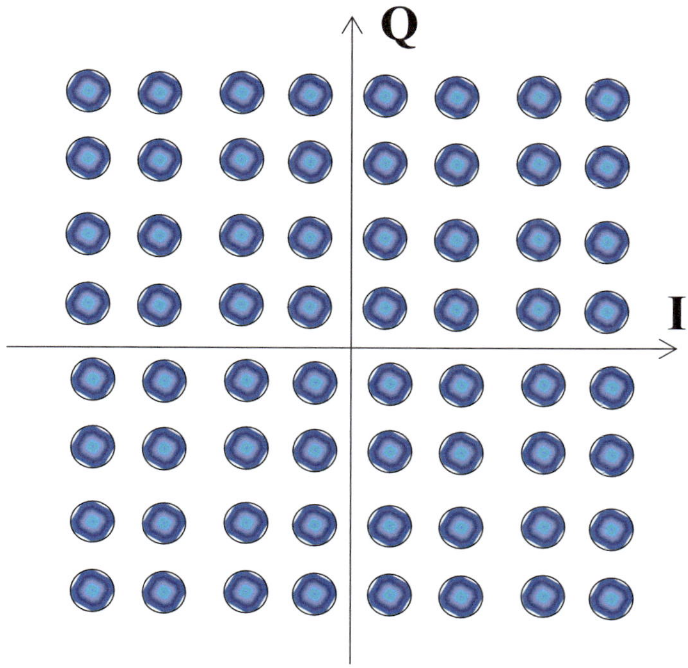

Fig. 3.9 Constellation of QAM signal (64QAM)

modulate the in-phase and quadrature components, followed by combining them with a phase difference of $\pi/2$ (as shown in Fig. 3.10b). For generating QAM signals, a Mach–Zehnder (MZ) modulator is typically used as the intensity modulator, and a dual-parallel MZ modulator, which integrates two MZ modulators in parallel, is commonly employed.

With the increase in modulation levels, the optical signal-to-noise ratio (OSNR) required to achieve the desired signal quality (bit error rate) also increases. Figure 3.11 shows the increase in the required OSNR due to multi-level modulation, assuming a constant bit rate (with the symbol rate reduced according to the modulation level). When comparing QAM to phase shift keying (PSK), where information is transmitted by only changing the phase states while keeping the amplitude of the optical signal constant, the increase in the required OSNR is smaller for QAM. However, even for QAM, when the information per symbol is doubled from 4QAM (QPSK) to 16QAM, the required OSNR increases by approximately 4 dB. This indicates that, under the same conditions for signal transmission, the spectral efficiency can be doubled, but the transmission distance is shortened to 0.4 times.

In early submarine cables, such as TPC-3, TPC-4, TPC-5, and China-US cables (1989–2000), NRZ-OOK was predominantly used. Subsequent 10 Gb/s wavelength division multiplexing systems, including PC-1, Japan-US, TGN-P, and UNITY (1999–2010), employed RZ-OOK and RZ-DPSK formats. Since 2016, coherent

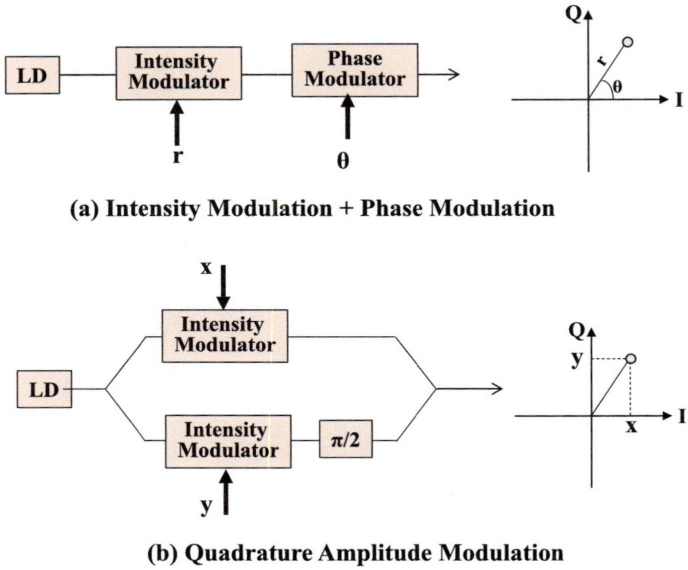

Fig. 3.10 Vector modulation

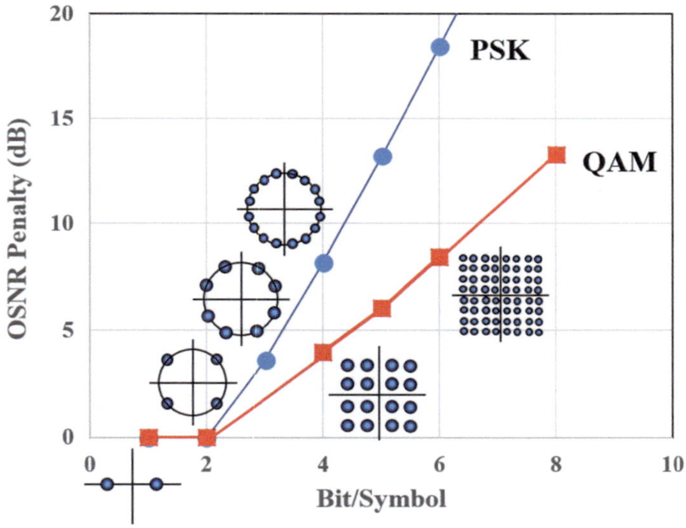

Fig. 3.11 Increase of required OSNR in higher level modulation format

reception techniques have been adopted, utilizing 4-level quadrature amplitude modulation (4QAM) in systems like the FASTER. The details of these systems are explained in the following chapters.

Problems

3.1 Draw the temporal waveforms of 10 Gb/s NRZ-OOK and RZ signals with a data pattern of "10110010."
3.2 Draw the temporal waveforms of 10 Gb/s PAM4 with the same data pattern as 3.1. Refer the data assignment for each pulse in Fig. 3.3.
3.3 Calculate the bit rate of DPSK and DQPSK signals with the time slot of 25 ps. Here, the time slot is the minimum time period in which the phase of light changes.
3.4 Draw the constellation diagram of 16PSK and 16QAM.
3.5 How many bits can be transmitted with one symbol with 16QAM and 64QAM?
3.6 Modulation format is changed from BPSK to QPSK and 8PSK with same average power and symbol rate. Calculate Euclidean distances of QPSK and 8PSK when Euclidean distance of BPSK is A.

Reference

1. J. Winzer, R.J. Essiambre, Advanced modulation formats for high-capacity optical transport network. IEEE J. Lightwave Technol. **24**, 4711–4728 (2006)

Chapter 4
Long-Haul Optical Fiber Transmission Using Regeneration Repeaters

Abstract The first-generation long-haul submarine cable systems using optical fiber employed regeneration repeaters. In a regeneration repeater, optical signals are converted to electrical signals and the distorted waveforms are cleaned with reshaping and retiming functions in the electrical circuits. Then the regenerated optical signals are launched into the next section. This chapter presents an overview of the optical regeneration repeater technologies and commercial Trans-Pacific submarine cable systems using optical regeneration repeaters.

Keywords Regeneration repeater · Reshaping · Regeneration · Retiming · Avalanche photo diode · 1.3 μm wavelength band · 1.55 μm wavelength band · Dispersion shifted fiber · TPC-3 · TPC-4

4.1 Overview of Optical Regeneration Repeater Transmission

In optical regeneration repeater systems, the transmission distance for optical signals becomes the interval between repeaters when designing the system. A typical example for the repeater spacing is approximately 50 km for TPC-3 using the 1.3 μm wavelength band and about 130 km for TPC-4 using the 1.5 μm wavelength band. Digital signals consisting of 0s and 1s are transmitted as optical signals.

Figure 4.1 shows the configuration of optical regeneration repeater system, where the bit error rate ε of each regeneration repeater section is typically designed to be a very small value, below 10^{-11} [1]. When n repeater sections are connected in series, the bit error rate (BER) becomes approximately n times the original rate.

$$\left\{1-(1-\varepsilon)^n\right\} \approx n\varepsilon \quad (\epsilon \ll 1) \tag{4.1.1}$$

In trans-oceanic optical submarine cable systems, the value of n becomes large, typically between several tens and 100. Thus, if the BER for a single repeater section is 10^{-11}, after 100 stages of transmission, the overall BER would reach 10^{-9}. This

M. Suzuki and I. Morita, *Long-Haul Optical Fiber Communication Systems*, Textbooks in Telecommunication Engineering, https://doi.org/10.1007/978-3-032-19472-5_4

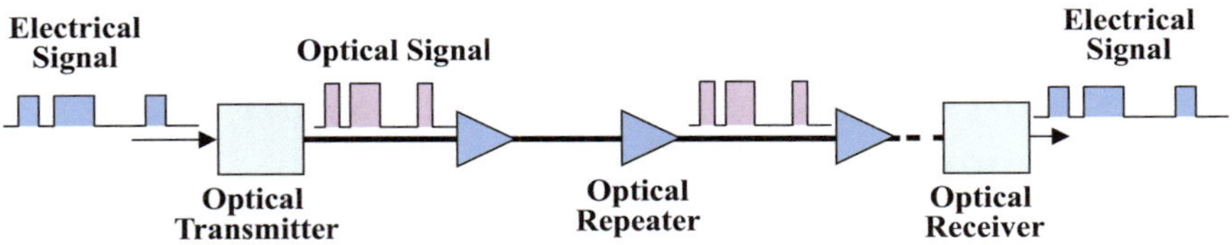

Fig. 4.1 Configuration of optical regeneration repeater system

error rate is generally acceptable for voice and video services but may cause issues for data services. Therefore, it is essential to achieve transmission quality that is nearly error-free.

4.2 Optical Regeneration Repeater

Figure 4.2 illustrates the basic configuration of an optical regeneration repeater. After transmission through an optical fiber, the optical signal is converted into an electrical signal by a photodiode (PD). Due to signal attenuation from fiber loss and waveform distortion caused by effects such as chromatic dispersion, an equalizing amplifier with low-pass characteristic is used first to amplify and filter out the fundamental frequency components of the data signal "1" (a process known as *reshaping*). The signal is then amplified by an automatic gain control amplifier to maintain a constant output and is decisioned by a decision circuit to determine whether it is 0 or 1. Timing is critical during this decision, and a synchronization circuit extracts the clock frequency and determines the phase within one cycle at which the decision is made (known as *retiming*). From the decision circuit, a rectangular wave pulse train consisting of 0s and 1s is regenerated (referred to as *regeneration*). This rectangular wave pulse train is amplified by a laser diode (LD) drive circuit and converted back into an optical pulse train by the LD.

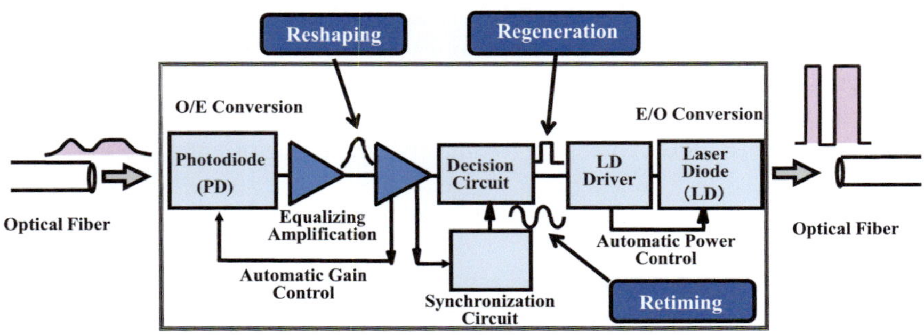

Fig. 4.2 Configuration of optical regeneration repeater

In regeneration repeaters, the characteristics of PD, which converts the optical signal into an electrical signal at the input, are crucial. Figure 4.3 shows the BER in relation to the received power for a PIN (p-type, intrinsic, n-type semiconductor)-PD, an avalanche PD (APD), and the shot noise limit, which represents ideal optical reception. Calculations assume a bit rate of 10 Gb/s, an NRZ (Non-Return-to-Zero) OOK (On-Off-Keying) modulation format, a temperature of 300 K, an internal quantum efficiency of 0.8 for both the PIN-PD and APD, an APD gain of 100, and an excess noise factor of 0.5.

For example, when using an APD, achieving BER below 10^{-9} requires a received power of approximately − 33 dBm, or 0.5 μW. This received power represents the average power. Assuming a mark rate of 50% for the pulse train consisting of 0s and 1s, the power of an optical pulse corresponding to 1 is 1 μW, as illustrated in Fig. 4.4. For a 10 Gb/s NRZ-OOK signal, the pulse width corresponding to 1 is 10^{-10} s (0.1 ns), so the pulse energy is 10^{-16} J, equivalent to 750 photons for light with a wavelength of 1.5 μm.

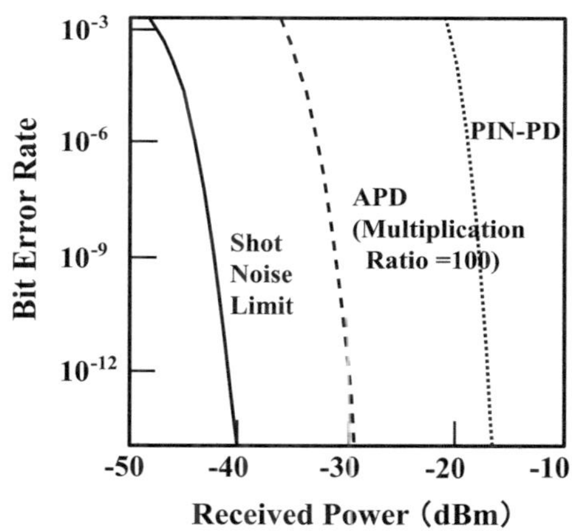

Fig. 4.3 BER characteristics with PIN-PD and avalanche PD (bit rate: 10 Gb/s, modulation format: NRZ-OOK, temperature: 300 K, internal quantum efficiency: 0.8, multiplication ratio and excess noise figure of APD: 100 and 0.5 are assumed)

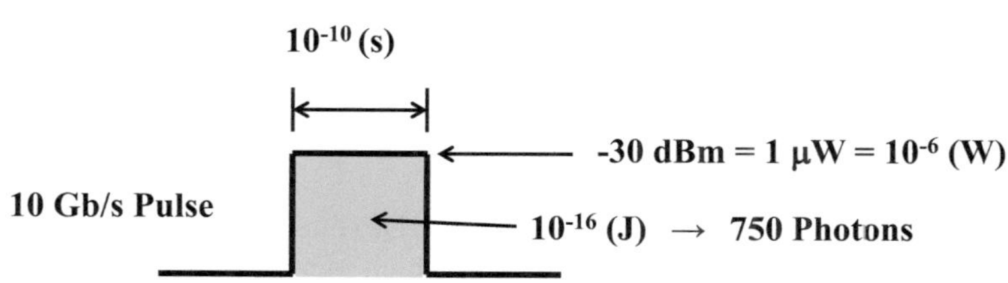

Fig. 4.4 Pulse energy for 10 Gb/s signal

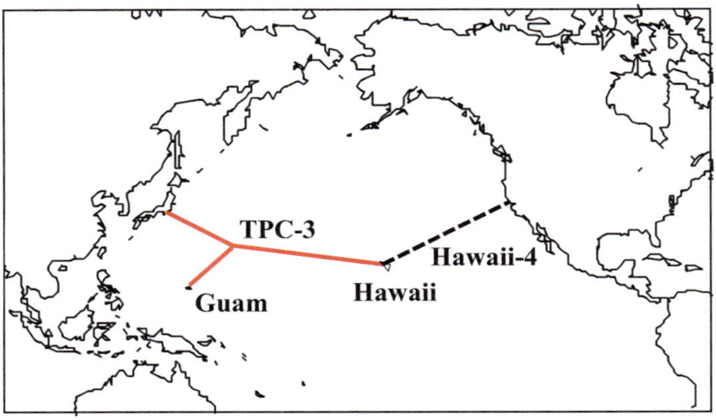

Fig. 4.5 The first optical submarine cable system between Japan and the United States, composed of TPC-3 linking Japan, Guam, and Hawaii, and Hawaii-4 connecting Hawaii to the U.S. mainland

4.3 Trans-Pacific Submarine Cable System Using 1.3 μm Regeneration Repeaters

The first optical submarine cable between Japan and the United States was composed of TPC-3 [1] and Hawaii-4, as shown in Fig. 4.5. The optical fiber used was the standard SMF operating at the 1.3 μm band, where chromatic dispersion is zero. Although the fiber loss in the 1.3 μm band is about 0.5 dB/km, which is higher than the 0.2 dB/km in the 1.5 μm band, it is the so-called zero-dispersion wavelength band, allowing the use of multi-wavelength emitting semiconductor lasers, such as Fabry-Perot (FP) lasers. The PD used was a Ge-based APD with a relatively high excess noise factor of 1, resulting in a repeater spacing of about 50 km. However, the transmission capacity per fiber was 280 Mb/s, offering a significantly broader bandwidth than previous coaxial cables, as illustrated in Fig. 1.1 in Chap. 1.

4.4 Trans-Pacific Submarine Cable System Using 1.5 μm Regeneration Repeaters

TPC-4 [2] shown in Fig. 4.6 used a single-mode DFB laser in the 1.5 μm band, the lowest-loss wavelength band for optical fiber, which minimized the impact of chromatic dispersion. This enabled a transmission capacity of 560 Mb/s per fiber, doubling the capacity, and significantly extended the repeater spacing to over 130 km. TPC-4 used the pure silica core fiber, which has even lower losses than the standard SMF with Ge-doped cores. In optical receiver, TPC-4 replaced the high-noise Ge-APD used in TPC-3, due to noise from avalanche gain, with a lower-noise InGaAs/InP-APD, further extending the repeater spacing. By reducing the number of repeaters

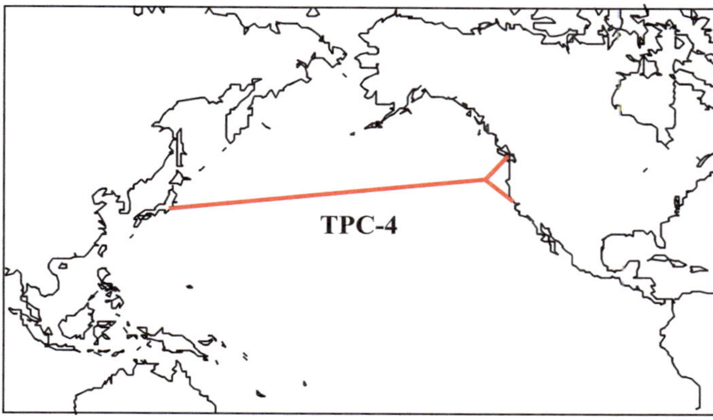

Fig. 4.6 TPC-4 directly connecting between Japan and USA, Japan and Canada

to less than half, a direct connection system between Japan and U.S.A. was realized, largely thanks to the advancements in optical fiber, single-mode laser and reception device characteristics, as covered in Chap. 2.

4.5 Challenges for Higher Speed Regeneration Repeater Transmission Systems

The transition from 280 Mbit/s in TPC-3 to 560 Mb/s in TPC-4 naturally led to scenarios targeting even higher speeds, such as 1 and 10 Gb/s. Various challenges also emerged, however, as shown in the configuration of the regeneration repeater in Fig. 4.2, it consists of numerous electronic components and circuits. As the bit rate increases, these electronic components and circuits must also be upgraded for higher speeds. Additionally, as shown in Fig. 4.4, a certain minimum number of photons per pulse is required to ensure error-free transmission. When the bit rate increases, the pulse width decreases, requiring higher optical power. The faster the electronic circuits, the greater the power consumption, and maintaining reliability with higher speeds also becomes an issue. These challenges brought optical amplification technology into the spotlight.

Problems

4.1 Calculate the required bit error rate (BER) in each span of a Trans-Pacific submarine cable system with the distance of 9000 km and the repeater spacing of 45 km assuming the required BER for the whole system is 1.0×10^{-12}.

4.2 Explain the 3R functions in the regenerator repeaters used in submarine cable systems.

4.3 Calculate the number of photons in an optical pulse corresponding “1” for 100 Gb/s NRZ-OOK transmission with average power of – 10 dBm and a mark ratio of 50%. Assuming the signal wavelength is 1.5 μm. Note that a photon energy E_p can be calculated as $E_p = hv$, here h is Planck constant, 6.6×10^{-34} J s, and v is frequency of light.

4.4 Explain the issues for increasing the bit rate to higher than 10 Gb/s in submarine cable systems with regeneration repeaters.

References

1. Y. Niiro, The OS-280M optical fiber submarine cable system. IEEE J. Lightwave Technol. **2**(6) (1984)
2. E. Nazuka, Y. Ogi, K. Shimizu, H. Homma, T. Tanabata, A. Nagai, Y. Yamazaki, H. Yamamoto, OS-560M optical submarine cable system, in *2nd International Conference on Optical Fiber Submarine Telecommunication Systems (Suboptic'93)*, Versailles, France, paper 8.2 (1993)

Chapter 5
Long-Haul Optical Fiber Transmission Using Optical Amplifier Repeaters

Abstract With the increasing traffic of long-distance communication, the need for high-speed communication with gigabit-per-second order has risen, driving the demand for innovations in transmission technology. The issues associated with increasing the bit rate in optical regeneration repeater systems became more apparent, leading to increased interest in optical amplification, which had been under investigation for some time. This chapter provides an overview of optical amplifiers and transmission systems using them in the repeaters. The first commercial Trans-Pacific submarine cable systems that adopted optical amplifier is also introduced.

Keywords Erbium-doped fiber amplifier · Noise figure · TPC-5 · Polarization dependence · Testbed · *Q*-factor · Coherent Optical Time Domain Reflectometry

5.1 Optical Amplifier

A significant breakthrough in optical amplification technologies was reported in 1989 at the Optical Fiber Communications Conference; transmission experiments using the Erbium-Doped Fiber amplifier (EDFA) pumped by a semiconductor laser were reported [1, 2]. The turning point for applying this technology to long-haul submarine cable systems came in the fall of 1989, with a report by KDD at the European Conference on Optical Communication on an experiment demonstrating optical amplification repeater transmission over 904 km using 12 EDFAs [3]. This expanded the typical 100 km regeneration section by about tenfold, proving that optical fiber system designs could address fiber loss challenges effectively.

Table 5.1 summarizes the characteristics of optical regeneration repeater systems and optical amplification repeater systems. In regeneration repeater systems, optical signals are regenerated through optical-electrical-optical conversion, requiring higher-speed electronic circuits and ICs with high reliability, which presents the significant drawback of needing specified components for each bit rate.

The optical amplification repeater, as shown in Fig. 5.1, involves doping the core of the fiber with erbium (Er) and pumped it with strong light at 1.48 μm or 0.98 μm to

M. Suzuki and I. Morita, *Long-Haul Optical Fiber Communication Systems*, Textbooks in Telecommunication Engineering, https://doi.org/10.1007/978-3-032-19472-5_5

Table 5.1 Comparison between regeneration and optical amplification repeater systems

Regenerator repeater	Optical amplifier repeater
• High speed analog and digital ICs • Dependent of modulation format/bit rate • Repeater span base design • Chromatic dispersion and nonlinear effects are reset at each repeater • BER is accumulated in the system • Polarization effects are reset at each repeater • A dedicated repeater is required for each channel	• Low speed analog ICs • Independent of modulation format/bit rate • End-to-end design • Chromatic dispersion and nonlinear effects are accumulated • Noise is accumulated • Polarization effects are accumulated • Multiple channels can be amplified simultaneously

utilize the EDFA for amplification at the 1.5 μm band. This approach has a very simple configuration and eliminates the need for high-speed electronic circuits specified for each bit rate, allowing for significant cost reductions in the repeaters.

On the other hand, optical amplifiers are accompanied by significant noise due to spontaneous emission. The noise characteristics are primarily determined by the noise figure, which can be approximated by the formula $2 \times n_2/(n_2 - n_1)$, where n_i is the number of electrons at N_i $(i = 1, 2)$. In the case of 1480 nm pumping, the signal wavelength of 1550 nm is close to the pump wavelength, making it difficult to achieve a sufficient population inversion where $n_1 \fallingdotseq 0$ due to the proximity to a two-level system consisting of N_2 and N_1. In contrast, 980 nm pumping forms a three-level system, allowing for adequate population inversion to be achieved, resulting in a noise figure that approaches the ideal value of 2 (or 3 dB). Figure 5.2 shows the typical input-output and noise figure characteristics of an EDFA. The graph assumes

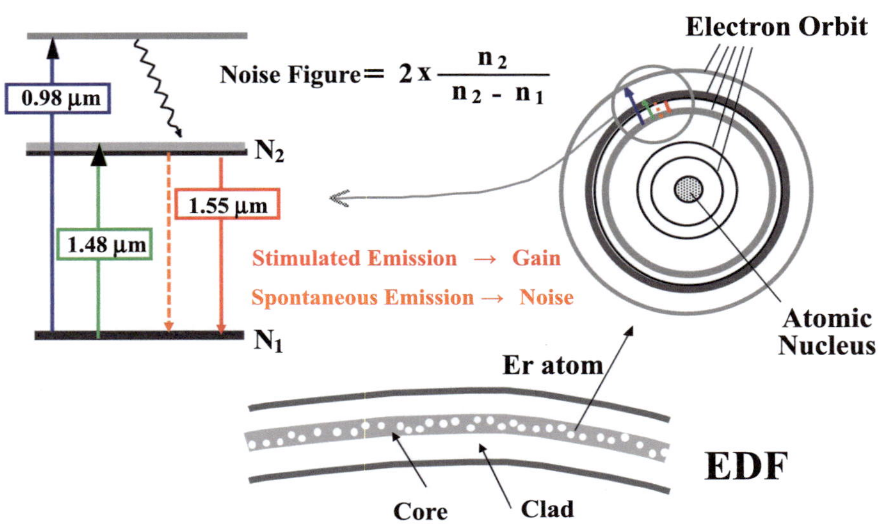

Fig. 5.1 Schematics of Er-doped fiber, electron orbits, and energy level

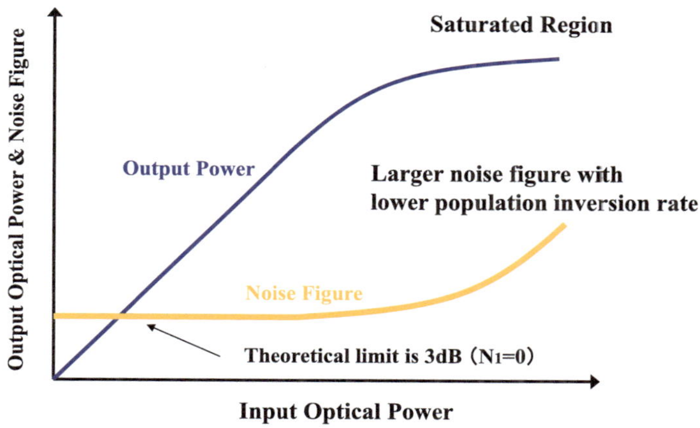

Fig. 5.2 Input-output characteristics and noise figure of optical amplifier

980 nm pumping, where the noise figure approaches the theoretical limit of 3 dB when the input signal is low. As the input signal increases, the output power saturates, causing the population inversion ratio to degrade with n_1 increasing, which in turn leads to an increase in the noise figure meaning the noise characteristics worsen. It should be noted that depending on the operating conditions, a difference of more than 3 dB in the noise figure can be observed between pumping with 1480 and 980 nm light. Furthermore, while the saturation characteristics of the output power may indicate degradation from a power output perspective, as will be discussed later, they contribute to output power stabilization in multi-stage amplification repeater systems.

5.2 Overview of Optical Amplifier Repeater Transmission Systems

Figure 5.3 shows a multi-stage optical amplifier repeater system with fixed intervals ranging from several tens of kilometers to 100 km. Since signal regeneration including waveform shaping and timing regeneration is not performed over long distances, issues such as optical noise, signal distortion, and timing jitter accumulate with distance.

Figure 5.4a schematically illustrates the wavelength dependence of the gain of an EDFA. With a single stage of the optical amplifier, a relatively wide range of wavelengths exceeding 30 nm is covered. However, as the number of stages increases, the gain becomes very narrow, as indicated by the dashed line. This phenomenon is similar to stacking bandpass filters and is referred to as self-filtering effect. To address this, as shown in Fig. 5.4b, by inserting a gain equalization filter that has a wavelength dependence opposite that of the EDFA characteristics, it becomes

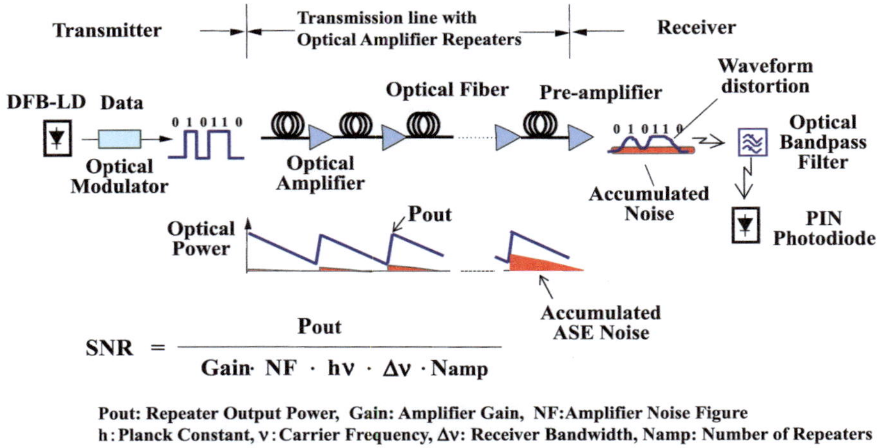

Fig. 5.3 Optical amplifier repeater system with concatenated multiple repeaters

possible to maintain a fairly broad amplification gain even when connected in multiple stages.

Figure 5.5 illustrates the circuit configuration of the optical amplifier, which incorporates various optical components in addition to the EDF. When the input optical signal, attenuated due to the loss in the optical fiber, enters the optical amplifier from the left, a portion of the light is first split off by the monitor coupler to monitor the power of the input signal using a photodiode (PD). After passing through the optical isolator, the signal light and pump light are combined using a WDM coupler and then incident on the EDF. The gain equalization filter, positioned after the output optical

Fig. 5.4 Wavelength characteristics of optical amplifier and gain equalization filter

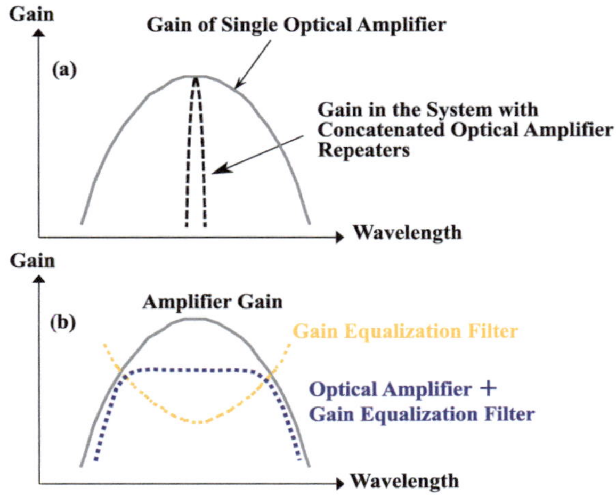

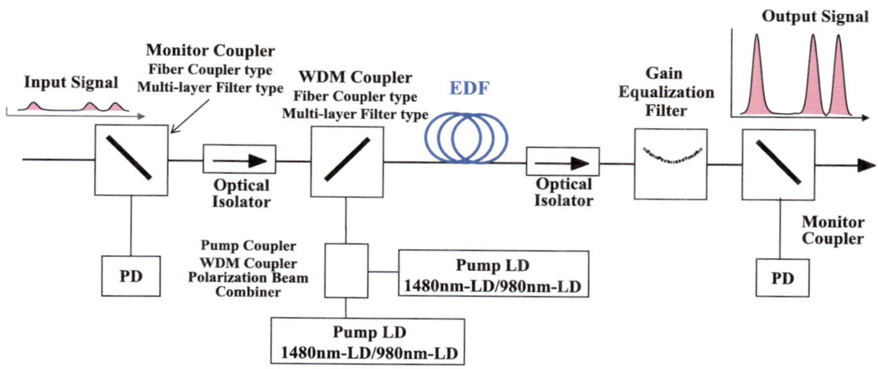

Fig. 5.5 Typical configuration of optical amplifier repeater

isolator, serves to flatten the wavelength characteristics, as shown in Fig. 5.4b. This filter is a critical component in WDM transmission, which is discussed in Chap. 6.

5.3 System Development

As an example of a long-distance optical amplification repeater system, we will discuss the system development aimed at realizing the Trans-Pacific Cable-5: TPC-5, which was the first Trans-Pacific optical amplification system. The target bit rate for this development was set at 5 Gb/s. This value was established in response to the increasing demand for communication traffic and is nearly ten times that of TPC-4 [4]. According to the bit rate hierarchy defined by Telecommunication Standardization Sector in International Telecommunication Union (ITU-T), the expected bit rates would typically be 2.5 Gb/s or 10 Gb/s. However, there was an understanding in submarine cable systems that the bit rate conversion could be performed at the landing stations to connect with terrestrial systems, meaning that the bit rates did not necessarily align with the hierarchical structure at that time.

Figure 5.6 shows the configuration of the optical amplifier being developed, which is simpler compared to the configuration in Fig. 5.5.

When considering the upstream transmission path is from Japan to the United States, the downstream path is from the United States to Japan. These two paths are referred to as a fiber pair. The configuration includes two 1480 nm lasers that pump the EDFs for the upstream and downstream paths, ensuring redundancy; if one laser fails, the gain decreases but optical amplification is still maintained. The configuration omits components such as input/output monitor couplers, PDs, input isolators, and gain equalization filters to simplify the design, especially since the number of optical amplifiers is large. This optical amplifier is then connected in a multi-stage configuration to create a system similar to that shown in Fig. 5.3.

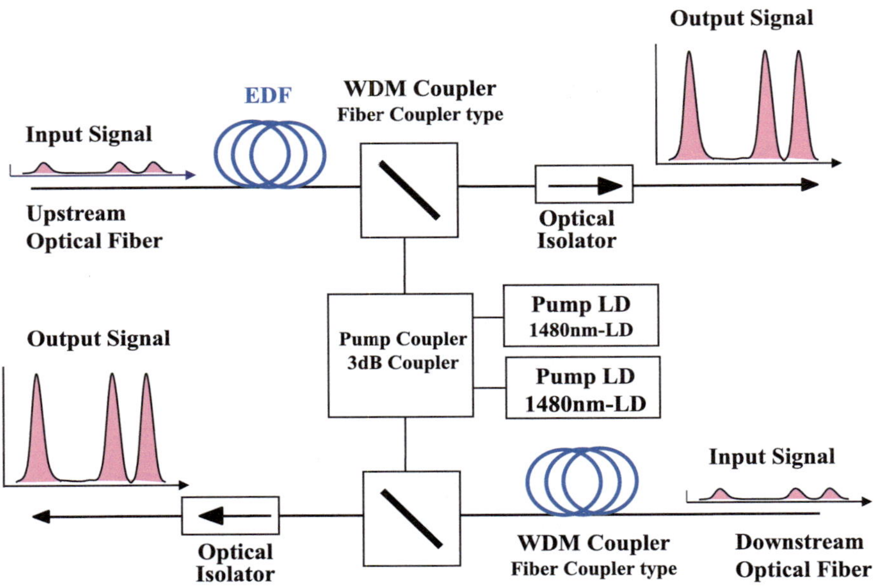

Fig. 5.6 Configuration of optical amplifier repeater developed for TPC-5

Optical components used in the amplifiers, such as optical isolators and WDM couplers, exhibit slight polarization dependence of typically around 0.1–0.3 dB, meaning that the optical amplification gain also has polarization dependence. Variations in the polarization state of the optical signal during transmission, due to temperature fluctuations or other factors, can lead to output power changes in the optical amplifier. As the number of repeaters increases, the output power fluctuations also become more pronounced. To mitigate this issue, a polarization scrambling technique was implemented to change the polarization state of the optical signal at the transmitter to avoid the formation of specific polarization states that would degrade performance. The actual system employed a low-speed polarization scrambler that changes the polarization state at around 50 kHz. Furthermore, because the cross-section of the optical fiber is not a perfect circle, the refractive index can vary slightly depending on the polarization state of the light signal, resulting in differences in signal propagation speed. This phenomenon is known as polarization mode dispersion (PMD), as discussed in Sect. 2.1. In an environment where the polarization state of the optical signal changes over time, the polarization dependence of the optical amplifiers and PMD can cause fluctuations in the signal-to-noise ratio, referred to as fading. For instance, Fig. 5.7 presents a conceptual diagram of the waveform when receiving the NRZ-OOK signals with a deta pattern of “010110.” This is fundamentally a statistical waveform similar to the waveform depicted in Fig. 2.27 of Sect. 2.4.

$$\mathrm{BER} = \frac{1}{2}\,\mathrm{erfc}\left(\frac{Q}{\sqrt{2}}\right) \approx \frac{\exp\left(-Q^2/2\right)}{Q\sqrt{2\pi}} \tag{5.3.1}$$

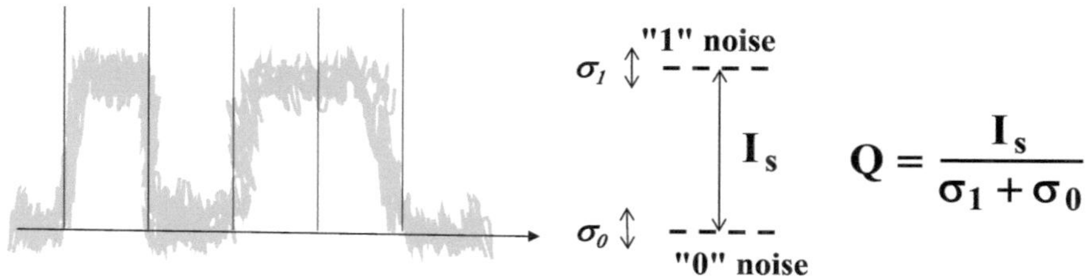

Fig. 5.7 Schematic of received optical signal

When the signal-to-noise ratio (SNR) is relatively high and waveform distortion is minimal, as shown in the figure, the distinctions between 0 and 1 can be clearly determined. However, if noise or distortion is significant, misjudgments can occur. The bit error rate (BER) in such case can be expressed by Eq. (5.3.1) (refer to Sect. 2.4). Under conditions with fading, the Q value fluctuates over time. A major focus during system development was how to estimate this fading. To understand the transmission characteristics of the signal passing through numerous optical components in a multi-stage optical amplifier system, initial experiments were conducted with recirculating loop where signals would circulate over optical fibers of several hundred kilometers to 1000 km (explained in detail in Chap. 6). Finally, large-scale transmission test equipment, known as a testbed, was used to simulate the actual Trans-Pacific distance of approximately 9000 km. The 9000 km testbed was collaboratively supplied by AT&T and KDD (currently KDDI), with each providing half of the distance, 4500 km each. This setup was installed at AT&T's research facility in Freehold during the three-year period from 1992 to 1995.

Figure 5.8 shows a photograph of the 4500 km segment of the testbed. On the front of the rack, there are knobs for adjusting the gain of the optical amplifiers which effectively adjust the output power of the pump lasers alongside LEDs displaying the set values. Behind them, the optical fibers that constitute the transmission path are housed. The repeater spacing was set to approximately 33 km, and the transmission fiber used was a dispersion-shifted fiber (see Sect. 2.1). While the development of the regeneration repeater system only required detailed evaluation of the repeater segment length, the development of long-distance optical amplification systems required test facilities that corresponded to the actual system length.

Using a 9000 km straight-line testbed close to the real environment, various practical development tests were conducted over a long period, including transmission tests, monitoring tests, and reliability tests [5]. In these tests, the characteristics of the system were repeatedly evaluated by changing the parameters. However, adjusting the output levels of the optical amplifiers required modifications to all approximately 270 amplifiers, making the setup process time-consuming. To simulate polarization variations in actual environments, various approaches were undertaken, such as altering the temperature of the optical fiber and changing the polarization of the

Fig. 5.8 Large-scale testbed used in system development

optical signal during transmission. Figure 5.9 schematically illustrates the fluctuations observed when measuring the Q value over long periods. Typically, measurements were conducted automatically during the night, with results compiled and analyzed the following morning; however, sometimes the measurements extended continuously for several days.

Evaluating system performance under fading conditions emerged as a crucial new point for long-distance optical amplification systems. Communication systems are fundamentally required to operate continuously, 24 h a day, 365 days a year, meaning any interruption in communication is unacceptable. However, achieving perfection

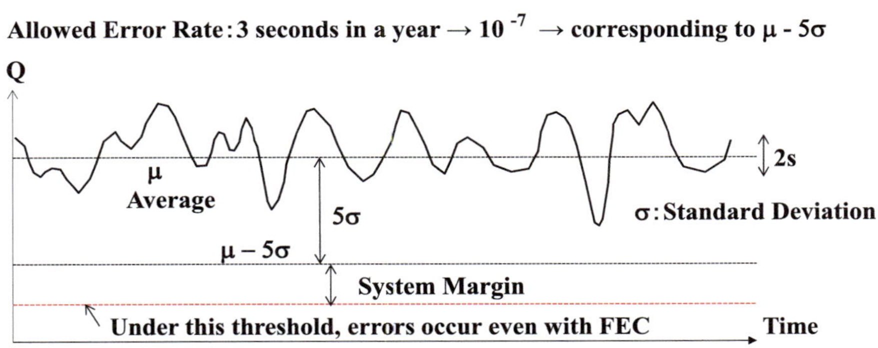

Fig. 5.9 Time variation of Q-factor corresponding to transmission performance

in real systems is impossible. Therefore, the target was set to ensure that the allowable downtime (error occurrence time) for long-distance communication systems did not exceed 3 s per year (10^{-7}). As shown in Fig. 5.9, when analyzing the long-term distribution of the Q value, it closely approximated a Maxwell–Boltzmann distribution, which is a slightly asymmetric normal distribution. The mean was denoted as μ, and the square root of the variance equivalent to the standard deviation was represented as σ. The goal was for $\mu - 5\sigma$ to correspond to a Q value exceeding the specified BER. The target BER was effectively set to be error-free (below 10^{-12}) or below 10^{-9}. The difference between the Q value and $\mu - 5\sigma$ represented the system margin in design.

5.4 Main Features of Optical Amplifier Repeater Transmission Systems

One of the features of long-distance optical amplification systems is their self-recovery characteristics in response to localized failures. Figure 5.10 illustrates this example, where a specific section of the fiber experiences an increase in loss due to some cause. As a result, the optical level decreases at that point, leading to a lower input optical level to the next stage of the optical amplifier than the designed value. In this situation, the optical amplifier exhibits larger gain due to its saturation characteristics, as shown in Fig. 5.11. Consequently, as the signal passes through several optical amplifiers, the signal level recovers. It is important to note that this recovery occurs only if the original settings are within the gain saturation region. Therefore, it can be said that most long-distance optical amplification systems operate in the gain saturation region. This self-recovery characteristic is crucial for enhancing the reliability of the system and ensuring the continuity of communication.

Next, we will discuss the coherent Optical Time Domain Reflectometry (OTDR) developed for application in long-distance optical amplification systems, which is used to measure the loss distribution of optical fibers and identify fault points [6]. OTDR, often referred to as an optical pulse tester, involves injecting a strong optical pulse into an optical fiber and measuring the distance dependence from the propagation time of the subsequent Rayleigh scattering and Fresnel reflections. Since the width of the optical pulse directly affects the distance resolution, short pulses are typically used. However, in long-distance optical amplification systems that pass through multiple optical amplifiers, short pulses are not useful since they cannot propagate far enough. Therefore, a method was devised where the amplitude of the optical signal is kept constant, and only the portion corresponding to the optical pulse is frequency shifted. This frequency-shifted part of the reflected light is then detected with high sensitivity using a coherent reception method.

Figure 5.12 illustrates this concept schematically. As shown in Fig. 5.12a, a 10 dB coupler is placed at the output section of the optical amplifier to propagate the backscattered light from the upstream optical fiber to the downstream optical

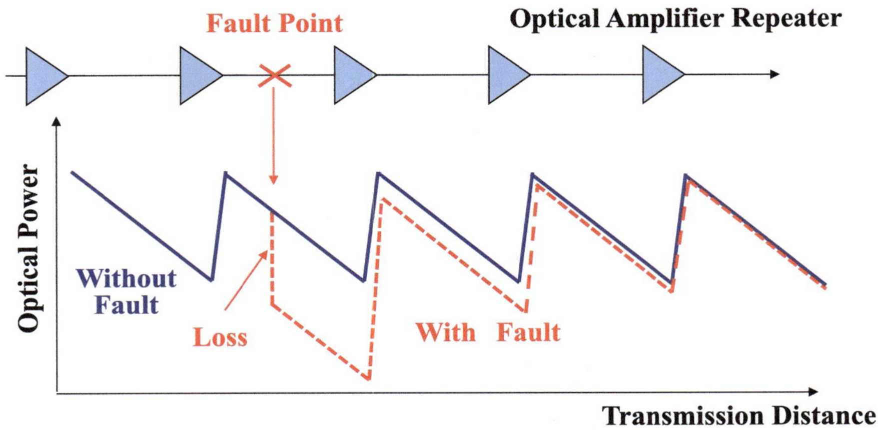

Fig. 5.10 Optical power level with a fault in optical amplifier system

Fig. 5.11 Power recovery in gain-saturation region of optical amplifier

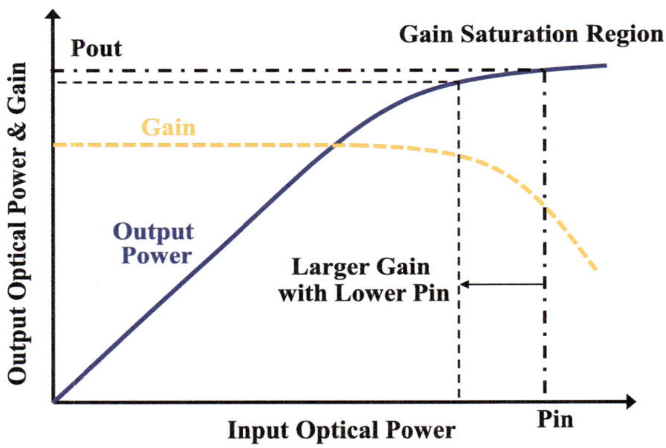

fiber. Since the backscattered light passes through two 10 dB couplers, it experiences a 20 dB loss. However, with high-sensitivity coherent reception, as depicted in Fig. 5.12b, the loss distribution of the optical fiber and local increases in loss can be measured across multiple stages. Figure 5.12c shows the measurement optical signal. The part corresponding to the short pulse of a conventional OTDR has its optical frequency shifted from the reference frequency f_0 by Δf. Heterodyne coherent detection is performed for the part corresponding to Δt, measuring the backscattered Rayleigh light from the optical fiber and the reflected light from local areas. T corresponds to the time equivalent to the measurement distance, which is 45 ms for 9000 km. To ensure distance resolution, Δt is set to a very small value relative to T. Therefore, advanced averaging processing allows for the measurement results depicted in Fig. 5.12b to be obtained.

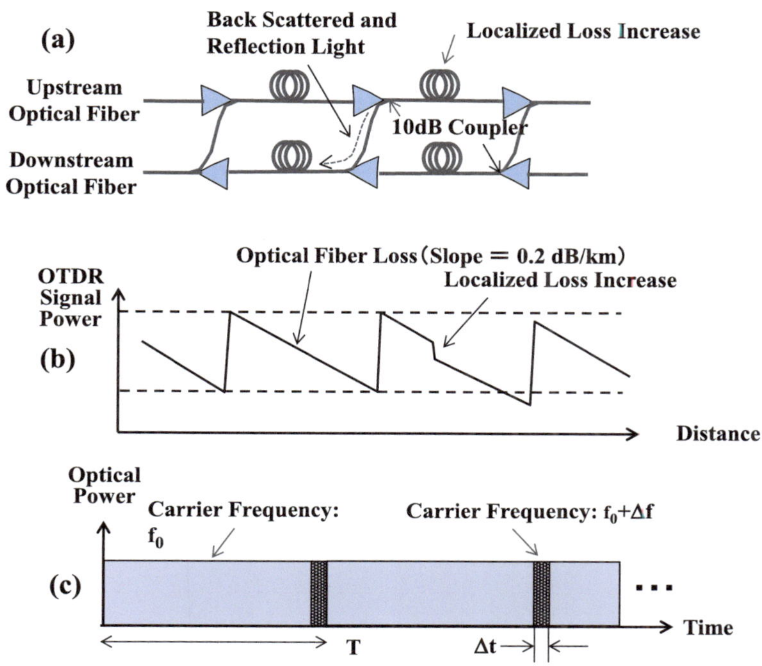

Fig. 5.12 Coherent OTDR in long-haul optical amplifier systems

5.5 First Trans-Pacific Submarine Cable System Using Optical Amplifier Repeaters

The TPC-5 cable network, utilizing optical amplification repeaters, was completed in 1996. As shown in Fig. 5.13, two cables were constructed in a ring configuration: a northern route that directly connects Japan and the United States across the Pacific Ocean and a southern route that passes through Guam and Hawaii, each serving as a backup for the other. The total length is approximately 25,000 km. The transmission capacity is 5 Gb/s, which is nearly nine times that of TPC-4, and later doubled through wavelength division multiplexing. The system consists of two fiber pairs, with each pair functioning as a ring network that backs up the other. There are two landing stations in Japan and the United States, designed to ensure the quickest recovery from failures or issues such as optical amplifier or fiber pair faults, or cable cuts, by activating protection actions.

The background for the initial ring configuration of the northern and southern routes lies in the fact that the large transmission capacity of 5 Gb/s could no longer be backed up by satellite communications, making it essential for the cable itself to have a backup system.

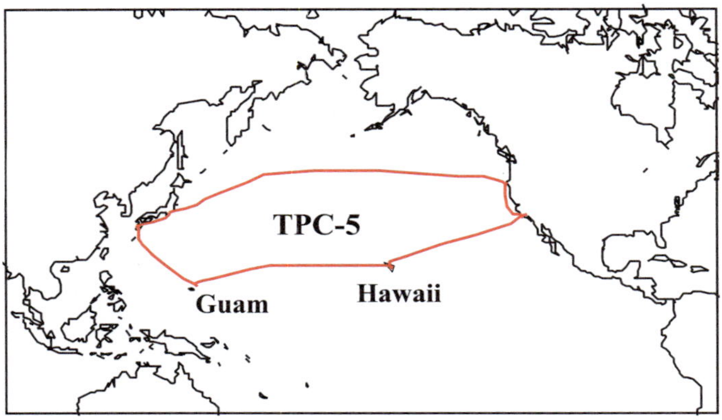

Fig. 5.13 TPC-5 cable network using optical amplifier repeaters

5.6 Further Development for Higher Capacity Optical Transmission Systems

A high-capacity submarine cable system capable of 5 Gb/s was realized, but around 1995, commercial internet traffic began to gradually increase, further enhancing the demand. This led to a focus on wavelength division multiplexing (WDM) transmission, the details of which will be discussed in Chap. 7, since it became clear that new developments were needed for significant performance improvements from the TPC-5 [7]. One aspect is the noise from the optical amplifier. TPC-5 utilized a 1480 nm pumped laser due to the development status of the pump laser and the required reliability. As mentioned in the explanation of Fig. 5.1, pumping at 1480 nm which is close to signal wavelength resulted in a large noise figure due to insufficient population inversion. However, using a 980 nm pump laser can improve the noise figure significantly. With 3 dB, improvement of the noise figure can double the number of repeaters or the capacity in multi-stage amplification systems, effectively extending the total system length or the capacity by a factor of two, which has a considerable impact on transoceanic systems. Another aspect is the optical fiber. Due to the development status of dispersion-shifted fiber, the effective area could not be made very large in TPC-5, making it susceptible to fiber nonlinear effects. By increasing the effective area of the fiber, the influence of nonlinear effects can be reduced, allowing for a higher fiber input power of the optical signal. Details on these topics will be covered in Chap. 6 and 7 regarding high-speed signal transmission and wavelength division multiplexing.

Problems

5.1 What are the two wavelengths of the pump light in EDFAs used for amplification of optical signal around 1.55 μm?

5.2 Explain the difference in the amplification characteristic of EDFA with two pump wavelengths in 5.1.

5.3 Calculate the signal-to-noise ratio (SNR) in the transmission system with the following parameters using the equation shown in Fig. 5.3 and assuming EDFA compensates for the fiber loss of each span (EDFA gain = Span loss).

- System length: 10,000 km.
- Signal wavelength: 1550 nm.
- Signal power: 0 dBm.
- Fiber attenuation: 0.2 dB/km.
- Repeater spacing: 50 km.
- Noise figure: 3 dB.
- Δv: 50 GHz.
- Planck constant: 6.6×10^{-34} J s.

5.4 Calculate the SNR when the repeater spacing is doubled to 100 km in the transmission system in 5.3.

References

1. M. Nakazawa, Y. Kimura, K. Suzuki, Soliton amplification and transmission with an Er^{3+}-doped fiber repeater pumped by InGaAsP laser diode, in *OFC'89, PD2*, 9 Feb 1989
2. H. Hagimoto, K. Iwatsuki, A. Takada, M. Nakazawa, M. Saruwatari, K. Aida, K. Nakagawa, M. Horiuchi, A 212 km non-repeated transmission experiment at 1.8 Gb/s using LD pumped Er^{3+}-doped fiber amplifiers, in *OFC'89, PD15*, 9 Feb 1989
3. N. Edagawa, Y. Yoshida, H. Taga, S. Yamamoto, K. Mochizuki, H. Wakabayashi, 904 km, 1.2 Gbit/s non-regenerative optical fibre transmission experiment using 12 Er-doped fibre amplifiers. Electron. Lett. **26**(1) (1990)
4. H. Taga, N. Edagawa, S. Yamamoto, S. Akiba, Recent progress in amplified undersea systems. IEEE J. Lightwave Technol. **13**(5), 829–840 (1995)
5. N.S. Bergano, C.R. Davidson, G.M. Homsey, D.J. Kalmus, P.R. Trischitta, J. Aspell, D.A. Gray, R.L. Maybach, S. Yamamoto, H. Taga, N. Edagawa, Y. Yoshida, Y. Horiuchi, T. Kawazawa, Y. Namihira, S. Akiba, 9000 km, 5 Gbit/s NRZ transmission experiment using 274 erbium-doped fiber-amplifiers, in *Optical Amplifiers and Their Applications. 3rd Topical Meeting, OAA '92*, Santa Fe, NM, paper PD11 (1992)
6. Y. Horiuchi, S. Ryu, K. Mochizuki, H. Wakabayashi, Novel coherent heterodyne optical time domain reflectometry for fault localization of optical amplifier submarine cable system. IEEE Photon. Technol. Lett. **2**(4) (1990)
7. S. Akiba, S. Yamamoto, WDM undersea cable network technology for 100Gb/s and beyond. Opt. Fiber Technol. **4**, 19–33 (1998)

Chapter 6
Dispersion Managed Optical Fiber Transmission Considering Fiber Nonlinearity

Abstract In transoceanic optical amplifier repeater systems with transmission distances reaching 10,000 km, various factors causing transmission performance degradation accumulate over distance. As a result, the chromatic dispersion and fiber nonlinearity lead to significant degradation of the signal waveform, making it difficult to achieve higher bit rates. After the development of the 5 Gb/s single-channel Trans-Pacific optical submarine cable system, TPC-5, significant efforts have been made to investigate high-speed transmission technologies to expand bit rates by 4–8 times. However, it was difficult to increase the bit rate to over 10 Gb/s with conventional rectangular NRZ-OOK optical signals in long-distance transmission over Trans-Pacific distance. This chapter discusses single-channel long-distance transmission using RZ-OOK optical signals suitable for high-speed transmission, targeting bit rates of 20–40 Gb/s.

Keywords Dispersion management · Optical soliton · Gordon–Haus timing jitter · Dispersion-managed soliton · Recirculating loop experiment · Chirped Gaussian-type Return-to-Zero optical pulse

6.1 10 Gb/s, 9,000 km NRZ-OOK Optical Signal Transmission

Figure 6.1 shows conceptual diagrams of NRZ-OOK optical signal waveforms and RZ-OOK optical signal waveforms. In NRZ-OOK optical signals, the optical pulse width varies based on the data pattern, since the light intensity remains constant over time in sections where "1" is continuous. On the other hand, in RZ-OOK optical signals, the optical waveform representing "1" has the same form regardless of the data pattern. Regarding fiber nonlinear effects, where the refractive index changes in response to optical intensity, RZ-OOK optical signals exhibit uniform behavior for all optical signals, making it simpler to manage fiber nonlinear effects compared to NRZ-OOK optical signals.

M. Suzuki and I. Morita, *Long-Haul Optical Fiber Communication Systems*, Textbooks in Telecommunication Engineering, https://doi.org/10.1007/978-3-032-19472-5_6

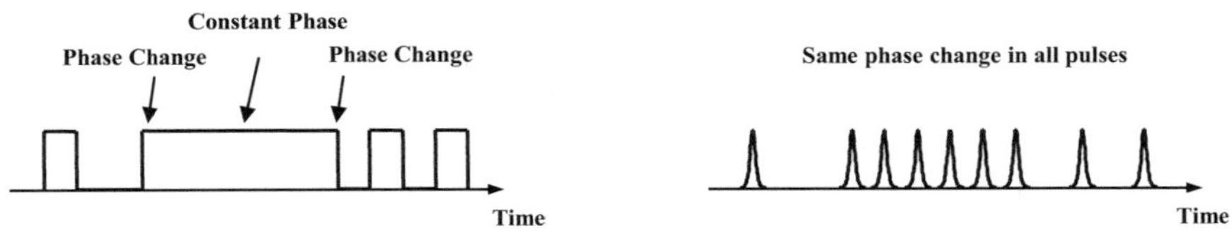

Fig. 6.1 Phase change by nonlinear effects in NRZ-OOK and RZ-OOK signals

Fig. 6.2 Waveform of 10 Gb/s NRZ-OOK signal after 9,000 km transmission

Figure 6.2 shows the received waveform of a 10 Gb/s NRZ-OOK optical signal using a 9,000 km straight-line testbed developed for TPC-5 [1]. In signal transmission at the zero-dispersion wavelength, the impacts of fiber nonlinearity become significant, resulting in transmission performance degradation. In this experiment, the signal wavelength was set slightly longer to achieve the best transmission performance with a bit error rate (BER) of less than 10^{-9}. It is noteworthy that the rectangular NRZ-OOK optical signal, during fiber transmission, experiences the fiber nonlinearity, leading to isolated pulses becoming RZ-shaped, and continuous pulses being compressed at their rising and falling edges. Since the advent of optical amplifiers, especially in long-distance optical fiber transmission systems such as Trans-Pacific links, optical fiber nonlinearity is unavoidable. Therefore, RZ-OOK optical signals, which can be uniformly controlled against fiber nonlinearity, can be considered a preferred signal format.

6.2 Optical Soliton Transmission

A representative method utilizing fiber nonlinearity with RZ optical signals is the optical soliton, which enables high-speed optical signal transmission without changing the signal waveform. It balances the pulse broadening due to the chromatic dispersion in the optical fiber, which has traditionally been considered a degradation factor for transmission performance, with the pulse compression caused by the fiber nonlinearity, enabling long-distance transmission without altering the optical signal waveform. The history of optical soliton dates back to its theoretical proposal in 1973 by Hasegawa et al. at AT&T as a stable solution to the nonlinear Schrödinger equation [2–4]. Subsequently, in 1980, Mollenauer et al. at AT&T first observed solitons in optical fibers, and in 1988, long-distance transmission over 6,000 km using Raman amplifiers was demonstrated. Experimental studies became more substantial in 1989 when Nakazawa et al. at NTT showed that in an optical amplification system using optical fibers with loss and EDFAs, it is possible to achieve nearly ideal optical soliton transmission by balancing the nonlinear effects and chromatic dispersion effects in the average power of the optical fiber sections [5, 6].

6.2.1 Optical Soliton

The wave equation considering the nonlinearity of lossless optical fibers is represented by the following nonlinear Schrödinger equation.

$$i\frac{\partial q}{\partial Z} + \frac{1}{2}\frac{\partial^2 q}{\partial T^2} + |q|^2 q = 0 \tag{6.2.1}$$

Here, q, Z, and T represent the normalized optical electric field, position, and time, respectively. The solution of the above equation is the optical soliton, and the optical soliton waveform is represented by the squared hyperbolic secant (sech^2), which is chirp-free (Fourier transform limit). Since Nakazawa et al. demonstrated the propagation theory of optical solitons in optical amplification systems where the loss of optical fiber is compensated by EDFA in 1989, active research has been conducted in various countries. The conditions for generating optical solitons in optical amplification systems can be expressed as follows:

$$P_{\mathrm{sol}} = \frac{0.776\lambda^3 A_{\mathrm{eff}} D}{\pi^2 c n_2 \tau^2} \tag{6.2.2}$$

Here, P_{sol} is the peak power of the optical soliton, A_{eff} is the effective area of the optical fiber (μm^2), $D(>0)$ is the chromatic dispersion (ps/nm/km), n_2 is the nonlinear coefficient (m^2/W), τ is the half-width at half-maximum of the optical pulse (ps), λ is the wavelength, and c is the speed of light. In optical amplifier repeater systems using Erbium-Doped Fiber Amplifiers (EDFAs), the distance between optical amplifiers, Z_{a}, needs to be set sufficiently small compared to the soliton period Z_0 as follows:

$$Z_0 = \frac{0.322\pi^2 c\tau^2}{\lambda^2 D} \tag{6.2.3}$$

In this case, the optical power launching to the optical fiber must be greater than P_{sol}, and the average power in the optical fiber segment needs to be set to satisfy the soliton condition. Optical solitons in optical amplification systems using EDFAs are referred to as dynamic solitons [6], path-average solitons [7], or guiding-center solitons [3, 4].

Optical solitons are the eigenvalues of the nonlinear propagation equation generated when all values of the chromatic dispersion D of the optical fiber, the parameter n_2/A_{eff} for fiber nonlinearity, the optical power level P_{sol}, and the optical pulse width τ of the sech^2-type optical waveform satisfy the soliton condition. However, even if various conditions deviate slightly, the optical pulse has a characteristic of naturally approaching that eigenvalue. If the conditions deviate significantly, dispersive waves, which are non-soliton components, occur, and the optical soliton disappears.

Figure 6.3 qualitatively explains the generation of solitons in optical fibers.

When the wavelength of the optical signal is longer than the zero-dispersion wavelength, the chromatic dispersion ($D > 0$) causes longer wavelengths (lower optical frequencies) to propagate more slowly, while shorter wavelengths (higher optical frequencies) propagate more quickly. An optical pulse consists of various frequency components determined by its pulse waveform, that is, the frequency components obtained by performing a Fourier transform on the pulse waveform. Therefore, when transmitting an optical pulse in an optical fiber, unless the wavelength of the optical signal is near the zero-dispersion wavelength, the optical pulse

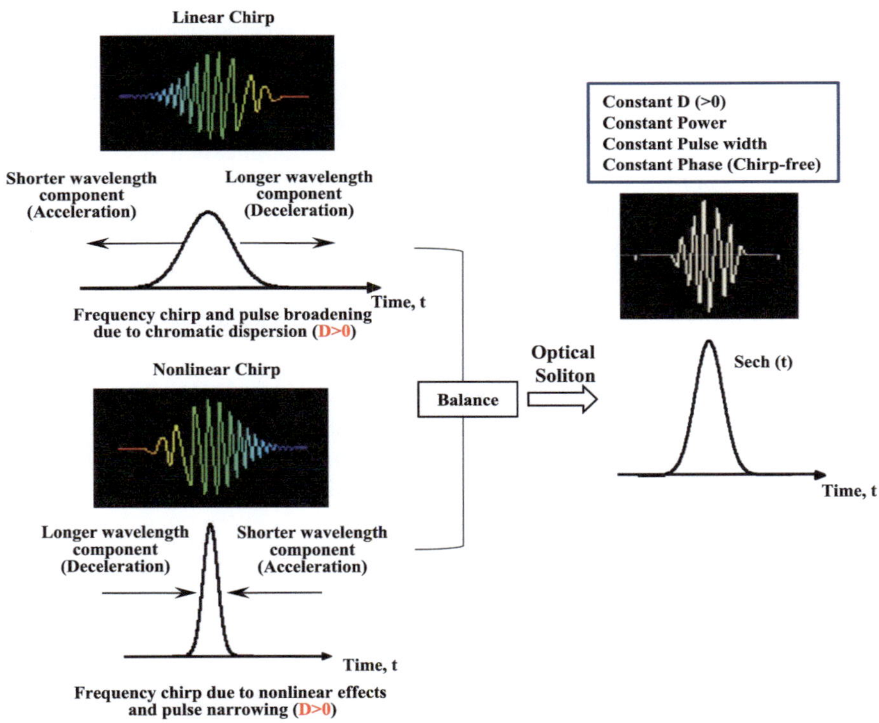

Fig. 6.3 Soliton transmission in optical fiber

will experience linear wavelength chirp due to the chromatic dispersion of the optical fiber, causing the pulse width to broaden. In the case of positive dispersion where optical solitons are formed, the chromatic dispersion causes the shorter wavelength components (higher frequency components) to accelerate while the longer wavelength components (lower frequency components) decelerate, leading to an increase in pulse width (see Fig. 6.3, top left). On the other hand, as discussed in Chap. 2, self-phase modulation (SPM), one of the fiber nonlinear effects, generates new longer wavelength components (lower frequency components) and shorter wavelength components (higher frequency components) as nonlinear wavelength chirps during the rising and falling edges of the optical pulse. With $D > 0$, the longer wavelength components (lower frequency components) in the rising edge are decelerated, and the shorter wavelength components (higher frequency components) in the falling edge are accelerated, resulting in the compression of the optical pulse (see Fig. 6.3, bottom left). Only when various parameters such as chromatic dispersion and optical waveform satisfy the conditions for soliton generation, the linear chirp be completely canceled out by the nonlinear chirp, allowing the optical pulse waveform to remain constant. Thus, a constant waveform with constant phase, chirp-free optical soliton is formed (see Fig. 6.3, right).

When applying optical solitons to communication systems, it is necessary to send an optical pulse for each data bit of "1." To suppress interference between adjacent pulses, the pulse width must be sufficiently small, less than one-fifth of the data interval. Typical examples of short-pulse light sources for optical soliton transmission include mode-locked lasers and fiber ring lasers, which can generate sequences of short optical pulses determined by the resonator length. On the other hand, Suzuki et al. proposed a compact, stable, and straightforward method that drives an electroabsorption (EA) modulator with a sinusoidal voltage [8]. Figure 6.4 illustrates the principle of generating short optical pulses using an EA modulator. The optical output of the EA modulator decreases approximately exponentially with respect to the voltage. When a laser beam of constant amplitude is incident on the EA modulator, and a deep DC bias voltage is applied to the EA modulator along with a sinusoidal modulation voltage, the optical output becomes short pulses as shown in Fig. 6.4. Due to the nonlinear transmittance characteristic of the EA modulator, the product of the optical output characteristics of the EA modulator and the sinusoidal function results in an intermediate waveform between Gaussian and sech^2 types.

Figure 6.5 shows the waveform and optical spectrum of the optical pulse generated by the EA modulator driven with a 10 GHz sinusoidal voltage. The optical pulse waveform is intermediate between Gaussian and sech^2 types, with a full width at half maximum (FWHM) of the pulse of 15 ps, sufficiently satisfying the condition of being less than one-fifth of the pulse interval of 100 ps. The FWHM of the spectrum is 23 GHz, and the time-bandwidth product is 0.34. This is close to the time-bandwidth product of 0.315 for a sech^2 pulse of the Fourier transform limit, indicating that excessive spectral broadening has not occurred. The pulse light source using a sinusoidally driven EA modulator can freely set the pulse width and repetition rate without being constrained by the optical resonator length, allowing for its use in many subsequent optical soliton transmission experiments.

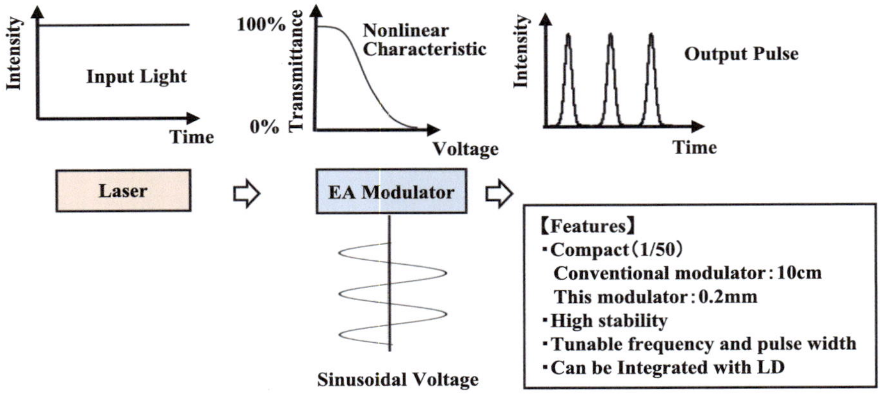

Fig. 6.4 Principle of soliton-pulse generation using EA modulator

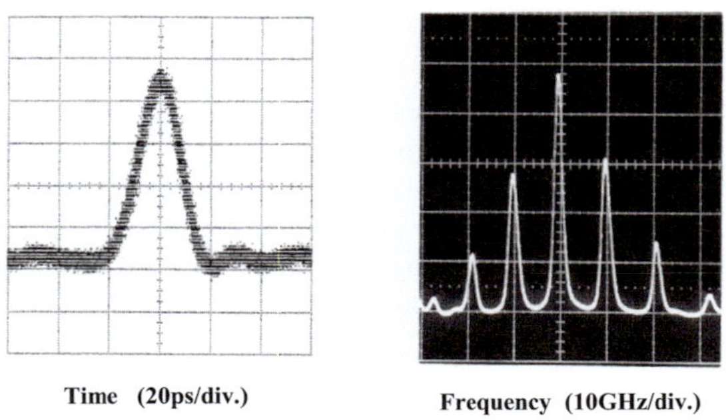

(a) Temporal waveform (FWHM: 15 ps) (b) Optical Spectrum (FWHM: 23 GHz)

Fig. 6.5 Optical pulse generation with sinusoidal-wave-driven EA modulator

6.2.2 Timing Jitter

Even with optical solitons that do not change shape in optical amplifier repeater systems, the transmission performance is significantly limited by the timing jitter of the received pulse train, fluctuations along the time axis of the optical pulse train, caused by the interaction of amplifier noise, fiber nonlinearity and chromatic dispersion.

(1) Gordon–Haus Timing Jitter

Fluctuations in optical intensity caused by ASE noise from the optical amplifiers are converted into carrier frequency fluctuations of the optical soliton through self-phase modulation (SPM) during the signal transmission over nonlinear fiber. These frequency fluctuations are then transformed into fluctuations in the arrival time of the optical pulses due to chromatic dispersion in the optical fiber. This phenomenon is referred to as Gordon–Haus timing jitter [9], which is the largest issue in optical soliton transmission using optical amplifiers. As shown below, Gordon–Haus jitter increases with the cube of the transmission distance L, making it a significant limiting factor for transmission distance.

$$\langle \delta\tau^2 \rangle = \frac{D^2 L^3}{Z_a} \langle \delta\omega^2 \rangle \tag{6.2.4}$$

Here, L is the transmission distance, Z_a is the amplifier spacing, $\langle \delta\tau^2 \rangle$ and $\langle \delta\omega^2 \rangle$ represent the variance of the arrival time fluctuations and center frequency fluctuations of the optical pulse, respectively,

(2) Timing Jitter Caused by Soliton–Soliton Interaction

When adjacent optical soliton pulses are in close proximity, the slight overlap of the tails of the optical pulses causes a minor change in both the intensity and the refractive index of the pulses. The jitter caused by soliton-soliton interference arises from the carrier frequency fluctuations due to this change in refractive index. By ensuring that the pulse interval is sufficiently wide, such as making it more than five times the pulse width, soliton–soliton interaction jitter can be reduced. However, caution is necessary as increasing the bit rate and shortening the pulse duration can lead to an increase in Gordon–Haus timing jitter.

6.2.3 Soliton Control

In long-distance optical soliton transmission, suppressing Gordon–Haus timing jitter, as explained in the previous section, is of utmost importance. As mentioned earlier, Gordon–Haus timing jitter is the result of fluctuations in the center frequency of the optical soliton caused by fiber nonlinearity, which are converted into timing fluctuations, represented as $\langle \delta\tau^2 \rangle$, after transmission through long-distance optical fibers with chromatic dispersion. Methods to suppress timing jitter in the frequency domain or time domain are referred to as soliton control. Below, we describe representative soliton control techniques.

(1) Sliding-Frequency Guiding Filter

By inserting optical bandpass filters into the transmission path, the fluctuation of the center frequency of the optical pulse can be reduced by making the loss slightly larger when the center frequency of the optical pulse deviates from the center frequency of the optical filters (see Fig. 6.6a). However, if fixed-wavelength optical bandpass filters are used, optical noise tends to accumulate near the center frequency of the filter. A method to avoid this is the sliding-frequency guiding filter method, which gradually changes the center frequency of the optical filters according to the transmission distance [10]. Since optical solitons continuously generate new frequency components due to self-phase modulation (SPM) during fiber transmission, they propagate while changing their center frequency in such a way that they become trapped by the sliding-frequency guiding filter. On the other hand, optical noise does not follow the change in center frequency and will eventually fall outside the bandwidth of the optical filters. This method is effective because it separates the optical noise and optical soliton signals in frequency domain. However, a practical issue with this method is the complexity in manufacturing the optical repeaters and maintaining the system, as the center frequencies of the optical filters equipped in each repeater must be different. Additionally, the difference between the input signal wavelength and the output signal wavelength complicates system design and management.

(2) Synchronous Intensity/Phase Modulation

By inserting optical intensity modulators or optical phase modulators into the transmission path and synchronizing the intensity/phase modulation with the original timing of the optical pulse train, a method called synchronous modulation can be used to return optical pulses with fluctuations deviating from the center position back to that center position [11]. In the case of intensity modulators, the synchronized optical modulators can directly suppress time fluctuations by forming optical gates that increase loss when slightly offset from the original pulse position (see Fig. 6.6b). The practical challenges of this method lie in the need for highly reliable ultra-fast optical modulators and high-speed electronic circuits in the inline optical modulation section, as well as the need to keep the ultra-fast modulation signal stably synchronized with the narrow input optical pulse (for example, a half-width of 10 ps $= 10^{-11}$ s).

The above methods are effective from the perspective of timing jitter suppression. With the sliding-frequency sliding guiding filter method, optical soliton signals at 10–20 Gb/s can be transmitted over tens of thousands of kilometers, while the synchronous intensity modulation method has been demonstrated to achieve stable optical soliton transmission over distances exceeding one million kilometers at 10 Gb/s and over tens of thousands of kilometers at 40 Gb/s.

However, as mentioned above, there are numerous practical challenges, and considerable sacrifices must be made to the advantage of the "simple repeater technology" that optical amplification repeater systems possess.

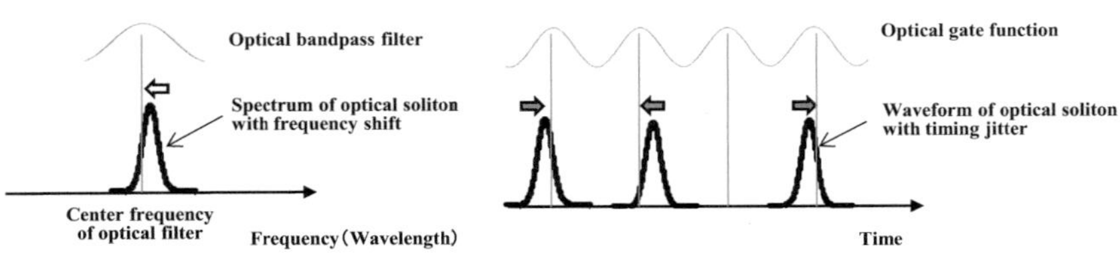

(a) Sliding-Frequency Guiding Filter (b) Synchronous Modulation

Fig. 6.6 Soliton control to suppress timing jitter

6.3 Dispersion Managed Soliton Transmission

Conventional timing jitter suppression methods focus on reducing the fluctuations in the frequency of optical signals or the timing jitter of optical pulse sequences either in the frequency or time domain, based on theory of optical soliton. On the other hand, focusing on the remaining parameter, chromatic dispersion, Suzuki et al. took a new approach that ensures the overall system has nearly zero chromatic dispersion through periodic dispersion compensation, so as to prevent the frequency fluctuations of the optical signal from being converted into timing jitter through the accumulated chromatic dispersion [12]. At first, there was skepticism on dispersion managed soliton; however, it was followed by many theoretical and experimental research worldwide and was proven that it was the newly founded periodically stationary solution of nonlinear Schrödinger equation of general dispersion-managed nonlinear fibers, and this method came to be known as "dispersion managed soliton."

For optical solitons to be established, an appropriate amount of chromatic dispersion is necessary. Therefore, it is not possible to reduce the chromatic dispersion of the transmission fiber to zero across the entire system. Instead, optical fibers that have a finite chromatic dispersion allowing for optical soliton transmission are selected, and optical fibers that compensate for the accumulated chromatic dispersion over distance are periodically inserted into the system. This configuration ensures that, locally, the conditions for optical soliton transmission are almost met, while macroscopically, the overall system exhibits nearly zero dispersion transmission. Figure 6.7a shows the dispersion map (the distance dependence of the accumulated chromatic dispersion) for the conventional optical soliton and dispersion-managed soliton.

As shown in Fig. 6.7a, in the conventional optical soliton, the accumulated chromatic dispersion DL increases with distance. Therefore, the fluctuations in the center frequency of the optical pulse caused by the noise of the optical amplifier and the fiber

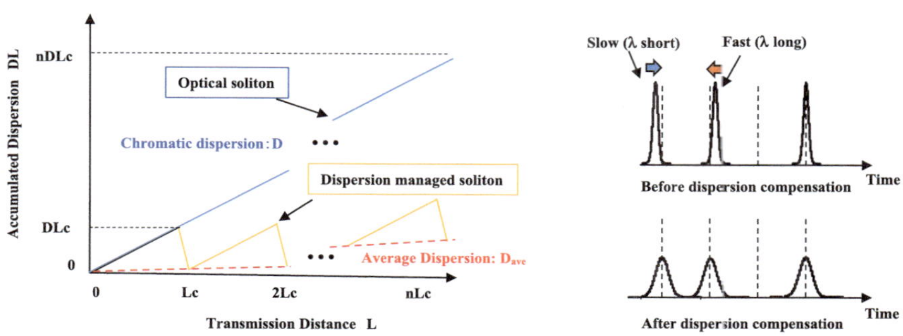

(a) Dispersion map

(b) Change of waveform and time position

Fig. 6.7 Dispersion map in conventional and dispersion-managed soliton system and optical waveform before and after dispersion compensation

nonlinearity are converted into significant timing jitter due to the increasing accumulated chromatic dispersion during long-distance transmission. On the other hand, in the dispersion-managed soliton, the accumulated chromatic dispersion is periodically offset to close to zero, making it less likely for timing jitter to be converted, even if frequency fluctuations are large. Figure 6.7b shows the changes in the position and waveform of the optical pulse after passing through the transmission fiber and after dispersion compensation. In the transmission fiber, the optical soliton propagates as a narrow pulse determined by the relationship between chromatic dispersion and fiber nonlinearity; however, as the center frequency shifts, the position of the optical pulse varies around a reference value. After dispersion compensation, since the accumulated chromatic dispersion is compensated, the position of the optical pulse returns to the center position, but the waveform broadens. When this broadened optical pulse enters the next transmission fiber, it is again compressed due to the fiber nonlinearity and chromatic dispersion. Unlike the conventional method that maintains the waveform of optical solitons, the dispersion-managed soliton is characterized by the periodic stretching and compression of the optical pulse, allowing for stable transmission. This enables stable optical transmission without waveform degradation or timing jitter in high-speed optical signal transmission.

In optical soliton communication, it was thought that large variations in chromatic dispersion would destroy optical solitons due to the massive generation of non-soliton components, as a constant chromatic dispersion is assumed. Until then, there were reports of demonstrations over a 3,000 km straight line that achieved an average chromatic dispersion value meeting the soliton condition by combining optical fibers with both positive and negative chromatic dispersions considering the variations in the characteristics of optical fibers [13], as well as computational simulation results [14]. However, there had been no attempts to perform large dispersion compensation sufficient to destroy solitons while reducing the average dispersion to less than about one-tenth of the soliton condition.

Next, we will discuss the interval for dispersion compensation, which is an important design parameter for a dispersion-managed soliton. Since stable optical pulses like those in conventional soliton transmission must be re-generated in the optical fiber transmission after dispersion compensation, there exists an optimal range for the dispersion compensation interval. In optical solitons, there is a distance called as soliton period Z_0 at which the nonlinear phase shift of the optical soliton equals 2π, serving as a guideline for the necessary transmission distance for stable soliton formation. To estimate the optimal dispersion compensation interval, computer simulations were conducted. In the case of a 20 Gb/s optical transmission system with a chromatic dispersion of $D = 0.2$ ps/km/nm and a pulse width of 7 ps, it was confirmed that compensating the accumulated dispersion approximately every 180 km (slightly less than twice the soliton period) results in a very good Q value of 19.8 dB after 9,000 km of transmission. If the dispersion compensation interval is set significantly shorter than the soliton period, it becomes almost equivalent to zero-dispersion transmission, leading to significant degradation of the transmission characteristics. Therefore, the dispersion compensation interval should be set to around the soliton period. This point represents a significant difference between dispersion compensation in the

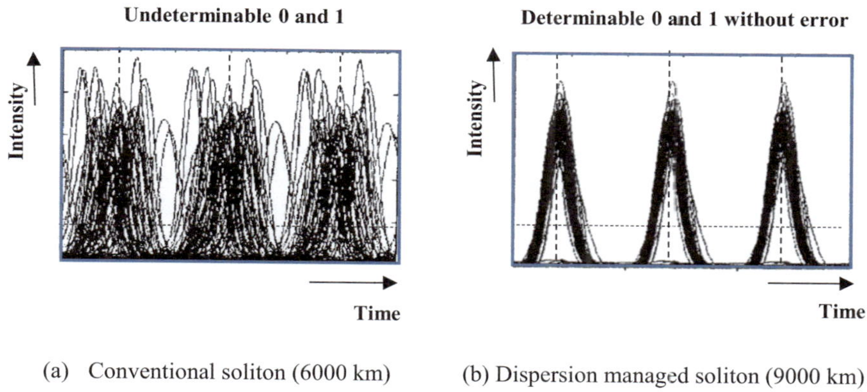

Fig. 6.8 Example of simulated received optical pulse (20 Gb/s)

systems considering fiber nonlinearity and linear systems. Figure 6.8 shows simulation results for the received waveform at 20 Gb/s for the conventional optical soliton with a uniform chromatic dispersion and the dispersion-managed soliton with periodic dispersion compensation. Figure 6.8a shows the waveform after 6,000 km of transmission for the conventional optical soliton. It can be seen that significant timing jitter makes decision of symbol, 1 or 0, nearly impossible. Figure 6.8b shows the waveform after 9,000 km of transmission for the dispersion-managed soliton (with Q = 19.8 dB). It is clear that there is almost no timing jitter or waveform degradation, indicating that stable transmission has been achieved.

6.4 System Design for 20 Gb/s Trans-Pacific Submarine Cable Systems

6.4.1 Recirculating Loop Experiments Using 1,000-km Fiber

To verify the applicability of the dispersion-managed soliton to Trans-Pacific submarine cable systems, optimal values for various parameters were sought using a recirculation loop employing 1,000 km of optical fiber. Figure 6.9 illustrates the recirculating loop experimental setup. On the optical transmission side, it consists of a DFB laser with an oscillation wavelength of 1557.8 nm, an EA modulator for generating RZ optical pulses with a pulse width of $\tau = 9$ ps, a $LiNbO_3$ data modulator operating at 10 Gb/s, a time-domain multiplexing circuit to double the bit rate of the short pulse sequence, a polarization scrambler operating at approximately 50 kHz, and a 10 GHz phase modulator for pre-chirping. The polarization scrambler was used to average out the polarization dependence of the optical amplifier gain and the polarization-dependent loss (PDL) of the components. Pre-chirping was employed to minimize the pulse width at the receiving end and optimize the transmission performance. The optical receiving side consists of a single-mode fiber (SMF) for residual dispersion compensation, an optical demultiplexer (EA modulator) that demultiplexes the 20 Gb/s time-division multiplexed signal into a 10 Gb/s signal, and a 10 Gb/s optical receiver. The 1,000 km optical fiber loop transmission path was composed of 30 spans of optical fiber with an average chromatic dispersion $D = 0.22$ ps/nm/km, 31 EDFA pumped at 1480 nm, 3 spans of dispersion-compensating fiber (DCF) for dispersion compensation every 10 spans (approximately 330 km) and optical bandpass filters (OBPFs) with 5 nm bandwidth. Since the soliton period Z_0 of the optical fiber transmission line was approximately 150 km, the dispersion compensation interval was set to about twice Z_0.

Figure 6.10 illustrates the time chart of the optical gate waveforms for optical switches 1 and 2, the optical input signal to the receiver, and the gate waveform for the bit error rate tester (BERT) required for the loop transmission experiment. The optical switch 1 on the transmitter side was turned on during the time T corresponding to the propagation time of the light in 1,000 km optical fiber, while the optical switch 2 in the loop transmission path was kept off during this time. Frequency-shift-free optical switches were used. After the optical signal has propagated for a time T through the loop transmission path, optical switch 1 was turned off, and optical switch 2 was turned on for a time nT. This allowed the incident light to propagate n loops within the transmission path. During this time, the optical signal that has completed its round trip through the loop transmission path was continuously sent to the receiver side through a 3 dB optical coupler. To evaluate the transmission performance after n rounds (after $n \times 1{,}000$ km of transmission), only the optical signal that has propagated n loops was gated by the BERT to measure the error rate in burst mode. By repeating this operation, the transmission performance could be evaluated through the recirculating-loop transmission experiment.

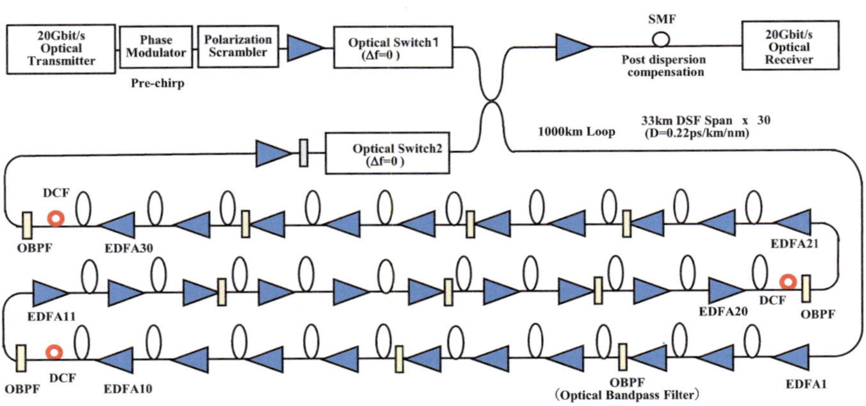

Fig. 6.9 Setup for 20 Gb/s transmission experiment using 1,000 km loop

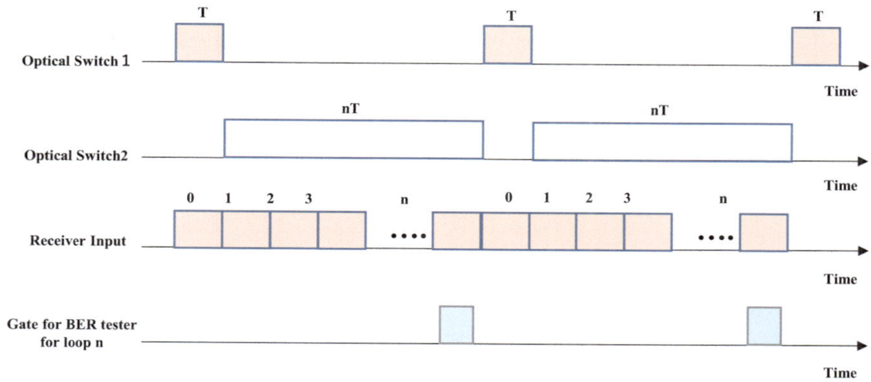

Fig. 6.10 Gate signals for optical switches and BER tester and receiver-input signal in recirculating-loop experiment

Figure 6.11a shows the relationship between the maximum transmission distance (the transmission distance at which the bit error rate is less than 1×10^{-9}) and the dispersion compensation ratio (compensated dispersion/cumulative chromatic dispersion) when optical bandpass filters were inserted into each optical amplifier. In the experiment, the pre-chirp on the transmitter side and the residual dispersion adjusted by single-mode fiber (SMF) on the receiver side were optimized to achieve the best transmission performance. The optimal dispersion for the SMF was approximately half the absolute value of the dispersion compensation fiber (DCF). From Fig. 6.11a, the optimal compensation ratio is estimated to be around 85%, with a margin of about ± 5%. The transmission performance improved by reducing the number of optical filters, so in the following tests, the number of optical filters was fixed at 12. Figure 6.11b shows the bit error rate characteristics with respect to the output level of the repeater. The evaluation of the bit error rate characteristics used the Q value. When the repeater output power is 1 dBm, the Q value is 19 dB (with a bit error rate of 2.5×10^{-19}), indicating that the Q value of 17 dB (with a bit error rate of 1×10^{-12}), which is a guideline for the transmission characteristics of the backbone transmission line, was comfortably achieved without forward error correction (FEC). Furthermore, from Fig. 6.11b, it can be seen that allowing for a 1 dB penalty provides a margin of approximately 2.5 dB for the repeater output power. It has already been established in systems like TPC-5 that fluctuations in repeater output power can be maintained at around 2 dB over 25 years, demonstrating that this transmission system has a sufficient power level margin for practical use. Additionally, the transmission distance achieving a bit error rate of 10^{-9} was 14,000 km, and it has been confirmed that when using low-noise EDFAs (noise figure < 4.0 dB) pumped by 980 nm laser diodes, the transmission distance with a bit error rate below 10^{-9} could be extended to 23,000 km.

Conventional optical soliton transmission has a small tolerance for power fluctuation because the optical power level that balances the fiber's dispersion is fixed. In

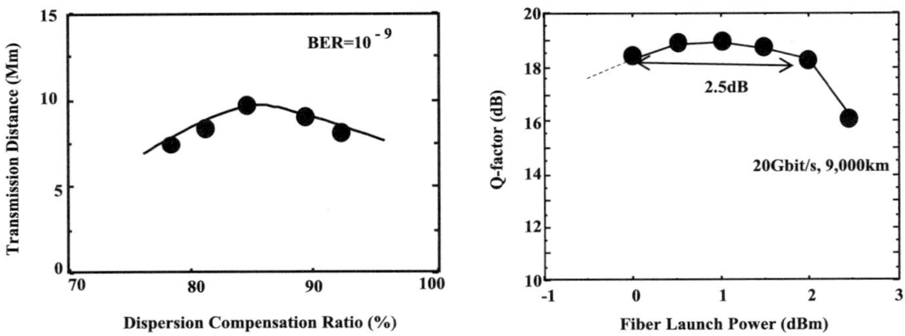

(a) Dispersion compensation ratio dependence (b) Fiber launch power dependence

Fig. 6.11 Transmission performance of 20 Gb/s dispersion-managed soliton

contrast, the dispersion-managed soliton significantly expands the tolerance for the power level. The substantial tolerance for chromatic dispersion and power level is also one of the essential characteristics that distinguish the dispersion-managed soliton from conventional optical soliton. The dispersion control, which sets the chromatic dispersion of the optical fiber in each part of the system to the desired characteristics, has already been adopted in existing optical submarine cable systems like TPC-5. Therefore, there are no practical obstacles to the implementation of periodic dispersion compensation methods.

6.4.2 8,123 km Straight-Line Experiments

To verify the practicality of the dispersion managed soliton, a large-scale straight-line transmission test of 8,123 km was conducted [15]. Figure 6.12 shows the experimental setup. The optical transmitter and the receiver had a configuration similar to that of the loop transmission test. The average dispersion compensation ratio of the 8,123 km optical fiber transmission line was fixed at 85% and the repeater power level was set at 1 dBm. The straight-line transmission system consisted of 8,123 km of DSF and SMF (hereinafter referred to as transmission fibers), 240 EDFAs, DCFs inserted every 10 repeaters for timing jitter suppression, and optical bandpass filters with a bandwidth of 3 nm inserted every 5 repeaters. The repeater spacing was set at 33–36 km, with an average amplifier gain of approximately 8 dB. Although the average chromatic dispersion of the transmission fibers was 0.2 ps/nm/km, the chromatic dispersion characteristics over the 8,123 km varied significantly, with over 40% of the total length being DSF with negative dispersion. The spans of the negative dispersion DSF were adjusted by adding SMF so that the average chromatic dispersion of each fiber span would be approximately 0.2 ps/nm/km. Since the chromatic dispersion of optical fibers in actual long-distance systems is not uniform and varies, this experiment was conducted in a transmission environment close to real conditions. Figure 6.13 shows the 20 Gb/s optical signal waveforms before and after the 8,123 km transmission. Even after the 8,123 km transmission, there was no significant degradation or timing jitter in the signal waveforms, indicating that good transmission was achieved.

Figure 6.14 shows the characteristics of the bit error rate (Q value) against transmission distance obtained from the straight-line transmission test and the 1,000 km loop transmission test.

The Q value after the 8,123 km transmission was approximately 19 dB, indicating that even in the straight-line test with significant variations in the optical fiber, good transmission performance was achieved similar to that of the loop transmission test, without the need for FEC. The degradation of the Q value with respect to transmission distance was linear, suggesting that the primary cause of the degradation in the Q value was the accumulation of ASE noise generated by the optical amplifier repeaters. The impact of timing jitter on degradation of transmission performance was negligible.

As part of the straight-line transmission experiment, a robustness test against repeater failures was conducted. The output power of the optical amplifiers within the repeater changes over time due to the gradual decline of the pump LD output during long-term use. Therefore, assuming a redundant configuration of the pump LDs, as shown in Fig. 6.15a, was used to simulate the failure of one pump LD, and the system's endurance against degradation of repeater output was evaluated by simulating the failure of one pump LD. Figure 6.15b shows the change in Q value with respect to the number of faulty repeaters. The results indicate that even with 13 failures (equivalent to a failure rate of approximately 5.4%), the degradation in Q

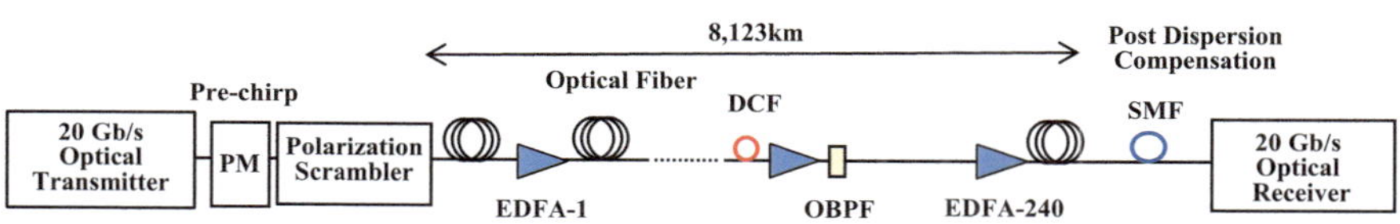

Fig. 6.12 Setup for 20 Gb/s, 8,123 km straight-line transmission experiment

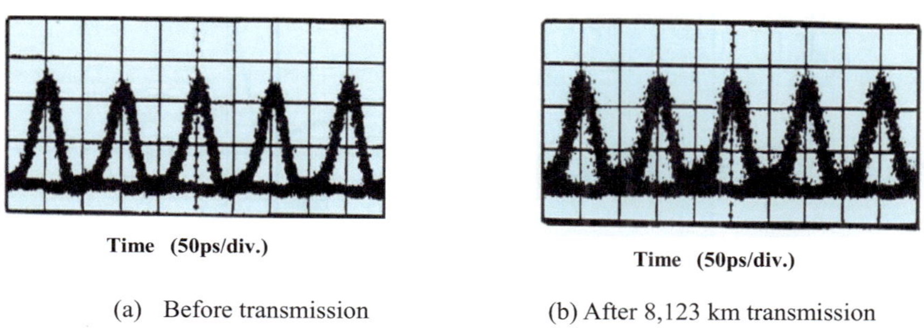

Fig. 6.13 Optical waveforms in 20 Gb/s transmission experiment

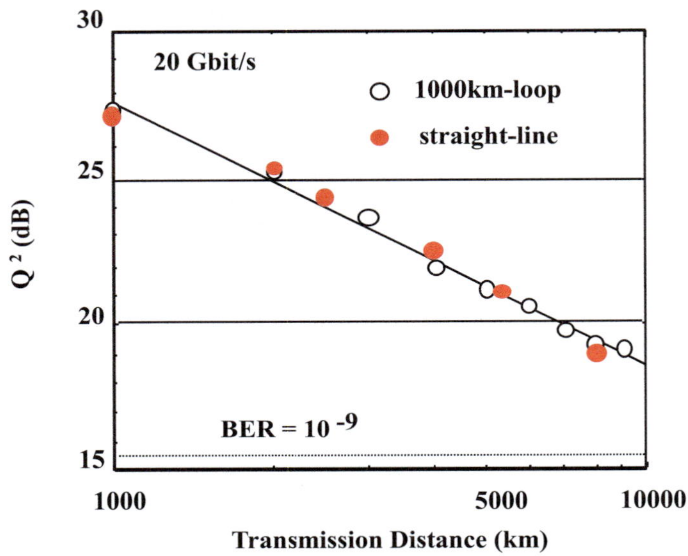

Fig. 6.14 Transmission performance of 20 Gb/s dispersion-managed soliton

value was small, approximately 0.7 dB. Since the TPC-5 was designed to maintain a failure occurrence rate of less than 1%, it has been confirmed that this transmission system has sufficient endurance against repeater degradation.

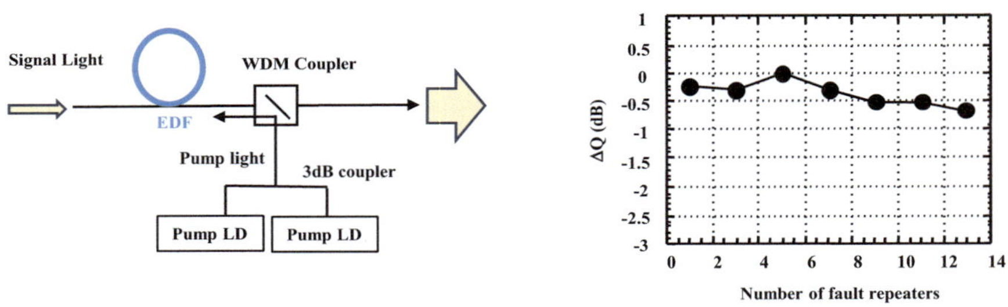

(a) EDFA with redundant pump LDs

(b) Q-factor degradation vs. number of fault repeaters

Fig. 6.15 *Q*-factor degradation by repeater faults

6.5 40 Gb/s, 10,000 km Dispersion Managed Soliton Transmission

In the previous section, it was shown that the dispersion-managed soliton provides transmission performance with sufficiently large margins applicable to Trans-Pacific systems at 20 Gb/s. Next, an evaluation of long-distance transmission performance at 40 Gb/s, which is eight times the speed of TPC-5, was conducted [16].

For high-speed signals at 40 Gb/s, the time slot per bit is 25 ps, and using a pulse width of 5 ps (which is one-fifth of the time slot) is not preferable, as it may lead to an increase in Gordon–Haus timing jitter. On the other hand, widening the pulse width degrades the transmission performance due to soliton–soliton interaction between adjacent soliton pulses. To solve this trade-off, the polarization between adjacent pulses was set to be orthogonal, and the transmitted optical pulse width was set to a relatively wide duration of 8–9 ps. By varying the chromatic dispersion of the 35 km optical fiber to $D = 0.18$, 0.31, and 0.78 ps/nm/km, and also changing the length of the dispersion compensation interval, the optimal dispersion map was investigated. The best transmission performance was obtained when $D = 0.31$ ps/nm/km, with a dispersion compensation interval of 70 km and an average dispersion of 0.028 ps/nm/km (dispersion compensation ratio: 93%). Figure 6.16 shows the distance dependence of the bit error rate from a 140 km recirculating-loop transmission experiment of a 40 Gb/s signal using single polarization and orthogonal polarization. In the case of single polarization, the transmission distance with a bit error rate below 10^{-9} was approximately 8,600 km, while with orthogonal polarization, the transmission distance extended to 10,200 km. Figure 6.17 shows the optical waveforms of the orthogonal polarization signal before and after 10,200 km of transmission. The adoption of orthogonal polarization suppressed soliton–soliton interaction, and by implementing optimal dispersion management, good transmission performance with no waveform degradation or timing jitter was achieved.

In this section, it was discussed that the dispersion-managed soliton, utilizing periodic dispersion compensation and RZ pulses in long-distance optical transmission systems, demonstrates extremely stable and excellent transmission performance. Additionally, as a single-channel system operating at bit rates of 20–40 Gb/s, which is 4–8 times faster than TPC-5, the dispersion-managed soliton method exhibits a high tolerance to various system parameter fluctuations required for Trans-Pacific long-distance optical transmission systems. Furthermore, it shows exceptional resilience to issues such as repeater failures, making it a highly practical transmission method not previously available.

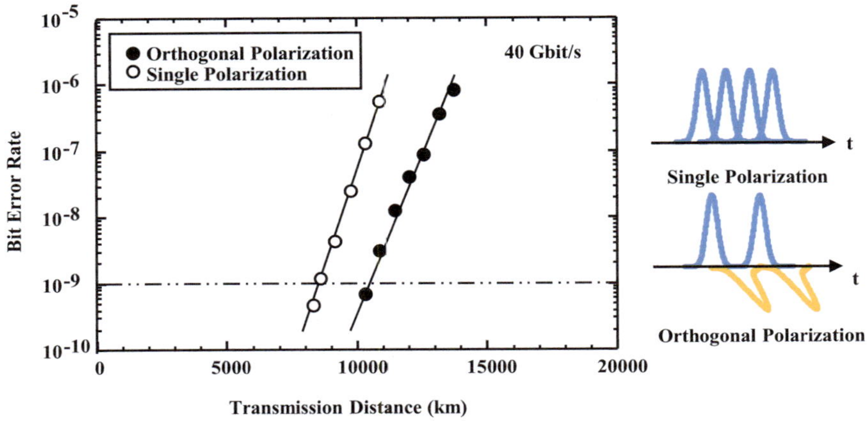

Fig. 6.16 BER in 40 Gb/s transmission with single and orthogonal polarization

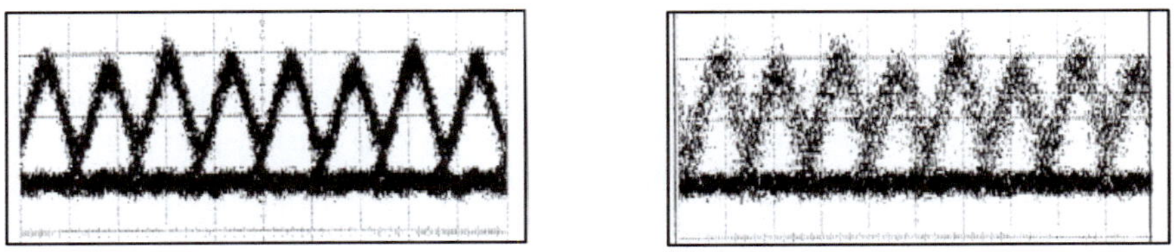

(a) Before transmission (b) After 10,200 km transmission

Fig. 6.17 Optical waveform in orthogonal polarization 40 Gb/s dispersion-managed soliton transmission

6.6 Generalization of Dispersion Management Considering RZ Optical Signals and Fiber Nonlinearity

So far, it has been shown that the dispersion-managed soliton, which employs RZ optical pulses and appropriate dispersion management, is a promising technology from the perspective of single-channel high-speed optical transmission systems. On the other hand, from the viewpoint of dramatically increasing the total capacity of the system, wavelength division multiplexing (WDM) is indispensable. In single-channel transmission systems, dispersion-shifted fibers (DSF) with low chromatic dispersion have been used as transmission fibers. However, in WDM systems, to suppress four-wave mixing (FWM) and cross-phase modulation (XPM), optical fibers with a relatively larger dispersion, such as non-zero dispersion-shifted fibers (NZ-DSF) and standard single-mode fibers (SMF), are employed. The dispersion-managed soliton has been generalized through subsequent theoretical studies [17–19], which have shown that the chirped Gaussian-type RZ optical pulse is a periodically stationary solution of a nonlinear Schrödinger equation of general dispersion-managed fiber consists of alternating positive and negative chromatic dispersion fibers.

Figure 6.18 illustrates the behavior of optical pulse waveforms in an optical transmission system using NZ-DSF with a dispersion value of − 2 ps/nm/km, compensated by positive dispersion SMF every 450 km. This dispersion arrangement is almost identical to the configuration of the Trans-Pacific submarine cable system with 10 Gb/s × 16 WDM channels, which will be discussed in the next chapter. In Fig. 6.18, negative dispersion NZ-DSF is placed at both the beginning and the end, and SMF is arranged in the center to compensate for the cumulative dispersion of both NZ-DSFs, resulting in an average dispersion of 0.2 ps/nm/km. This average dispersion value is a typical value used in standard optical solitons. The figure shows the propagation of optical pulses through this transmission path, as well as the changes in phase and amplitude during the pulse propagation. From Fig. 6.18, it can be seen that the pulse width of the optical signal repeatedly expands and shrinks within the section. In the absence of fiber nonlinearity, the waveform returns to its original shape due to dispersion compensation, resulting in the amplitude-phase trajectory tracing along a curve. However, in a nonlinear system, due to the influence of nonlinear chirp, the waveform does not return to its original shape after dispersion compensation, but rather, it takes on a symmetrical form after passing through the subsequent optical fiber transmission path. The motion on the amplitude-phase plane at this time traces a crescent-shaped trajectory, as shown in the figure. As illustrated, the optical pulse returns to a periodic stable point only when the pulse shape is a Gaussian waveform in RZ form.

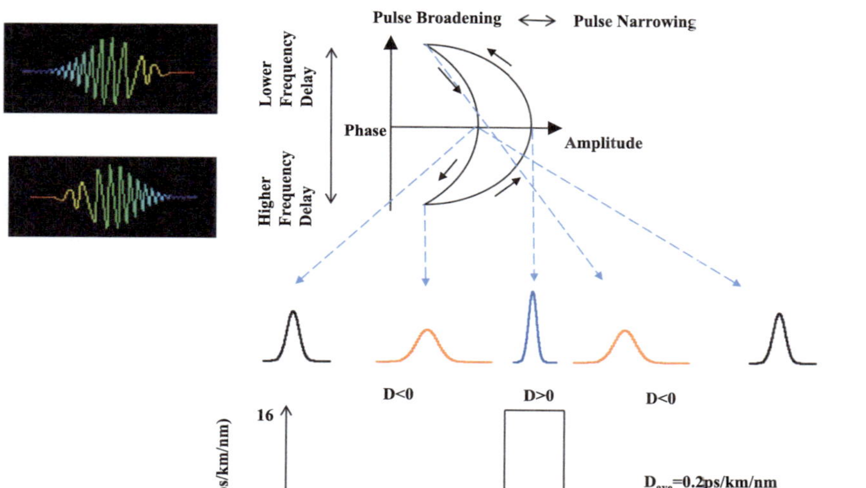

Fig. 6.18 Behavior of optical pulse waveforms in optical transmission system using NZ-DSF and SMF

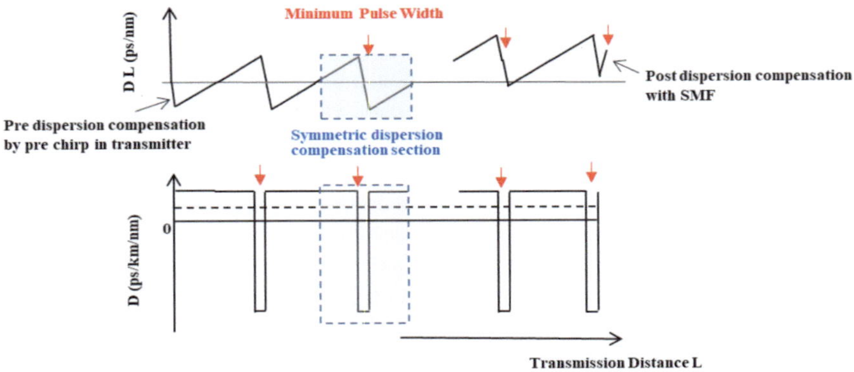

Fig. 6.19 Dispersion map considering pre- and post-dispersion compensation and chromatic dispersion of each optical fiber in 20 Gb/s dispersion-managed soliton transmission experiment

Although the optical waveform within the basic interval of dispersion compensation is changed, the optical waveform at periodic observation points is the same. In the optical soliton using optical fibers with a constant chromatic dispersion, the solution is a phase and amplitude constant sech^2-type optical pulse. In contrast, in the dispersion-managed soliton using optical fibers with periodically differing signs of chromatic dispersion, the steady-state solution is a chirped Gaussian-type RZ optical pulse that propagates while changing both phase and amplitude.

At the end of this chapter, we describe the relationship between the dispersion management used in the experiments and the symmetric dispersion management shown in Fig. 6.18. Figure 6.19 illustrates the dispersion map used in the 20 Gb/s Trans-Pacific transmission experiment, with pre-chirping and post-dispersion compensation, along with the periodic chromatic dispersion compensation for each dispersion-compensation section. Due to the pre-chirping at the transmitting, approximately half of the chromatic dispersion of the initial positive dispersion fiber is compensated. At the receiving end, the last DCF is compensated by SMF to be half of its value. Therefore, the basic interval of dispersion compensation (dashed line) becomes the symmetric dispersion management interval with opposite signs compared to Fig. 6.18, allowing for the stable propagation of Gaussian pulses.

Problems

6.1 Explain the condition regarding the parameters of optical fiber for optical soliton transmission.

6.2 Explain two soliton control methods for reducing the timing jitter in the conventional soliton transmission with a uniform chromatic dispersion.

6.3 Draw the dispersion map such as Fig. 6.7 for dispersion-managed soliton transmission using the positive and negative dispersion fiber with the following parameters.

- Chromatic dispersion of the positive dispersion fiber: 1 ps/nm/km.
- Chromatic dispersion of the negative dispersion fiber: − 90 ps/nm/km.
- Dispersion compensation interval: 180 km.
- Average chromatic dispersion of the transmission line: 0.2 ps/nm/km.

References

1. H. Taga, N. Edagawa, H. Tanaka, M. Suzuki, S. Yamamoto, H. Wakabayashi, N.S. Bergano, C.R. Davidson, G.M. Homsey, D.J. Kalmus, P.R. Trischitta, D.A. Gray, R.L. Maybach, 10 Gbit/s, 9,000 km IM-DD transmission experiments using 274 Er-doped fiber amplifier repeaters, in *Optical Fiber Communication Conference OFC'93, PD1*, San Jose (1993)
2. A. Hasegawa, F. Tappert, Transmission of stationary optical pulse in dispersive dielectric fibers, I anomalous dispersion. Appl. Phys. Lett. **23**, 142–144 (1973)
3. A. Hasegawa, *Optical Solitons in Fibers* (Springer-Verlag, 1990)
4. A. Hasegawa, Y. Kodama, *Solitons in Optical Communication* (Oxford University Press, 1995)
5. M. Nakazawa, K. Kimura, K. Suzuki, Soliton amplification and transmission with an Er^{3+}-doped fiber repeater pumped by InGaAsP laser diode. Electron. Lett. **25**, 199–200 (1989)
6. M. Nakazawa, K. Suzuki, H. Kubota, E. Yamada, Y. Kimura, Dynamic optical soliton communication. IEEE J. Quantum Electron. **26**, 2095–2120 (1990)
7. L.F. Mollenauer, S.G. Evangelides, H.A. Haus, Long-distance soliton propagation using lumped amplifiers and dispersion shifted fibers. IEEE J. Lightwave Technol. **9**, 194–197 (1991)
8. M. Suzuki, H. Tanaka, N. Edagawa, K. Utaka, Y. Matsushima, Transform-limited optical pulse generation up to 20-GHz repetition rate by a sinusoidally driven InGaAsP electroabsorption modulator. IEEE J. Lightwave Technol. **11**, 468–473 (1993)
9. J.P. Gordon, H.A. Haus, Random walk of coherently amplified solitons in optical fiber transmission. Opt. Lett. **11**, 665–667 (1986)
10. L.F. Mollenauer, J.P. Gordon, S.G. Evangelides, The sliding-frequency guiding filter: an improvement form of soliton jitter control. Opt. Lett. **18**, 897–899 (1993)
11. M. Nakazawa, K. Suzuki, H. Kubota, E. Yamada, Y. Kimura, M. Takaya, Experimental demonstration of soliton data transmission over unlimited distance with soliton control in time and frequency domain. Electron. Lett. **29**, 729–730 (1993)
12. M. Suzuki, I. Morita, N. Edagawa, S. Yamamoto, H. Taga, S. Akiba, Reduction of Gordon–Haus timing jitter by periodic dispersion compensation in soliton transmission. Electron. Lett. **31**, 2027–2029 (1995)
13. H. Taga, M. Suzuki, N. Edagawa, Y. Yoshida, S. Yamamoto, S. Akiba, H. Wakabayashi, 5 Gbit/s optical soliton transmission experiment over 3000 km employing 91 cascaded Er-doped fibre amplifier repeaters. Electron. Lett. **28**, 2247–2248 (1992)
14. M. Nakazawa, H. Kubota, Optical soliton communication in a positive and negative dispersion allocated optical fibre transmission line. Electron. Lett. **31**, 216–217 (1995)
15. N. Edagawa, I. Morita, M. Suzuki, S. Yamamoto, H. Taga, S. Akiba, 20 Gbit/s, 8100 km straight-line single-channel soliton-based RZ transmission experiment using periodic dispersion compensation, in *ECOC'95*, Brussels, Belgium, post-deadline paper, Th.A.3.5 (1995), pp. 983–986
16. I. Morita, K. Tanaka, N. Edagawa, M. Suzuki, 40 Gb/s single-channel soliton transmission over transoceanic distance by reducing Gordon–Haus timing jitter and soliton–soliton interaction. J. Lightwave Technol. **17**, 2506–2511 (1999)
17. N.J. Smith, W. Forysiak, N.J. Doran, Reduced Gordon–Haus jitter due to enhanced power solitons in strongly dispersion managed systems. Electron. Lett. **32**, 2085–2086 (1996)
18. Y. Kodama, Nonlinear chirped RZ and NRZ pulse in optical transmission line, in *Soliton Symposium*, Kyoto (1997), pp. 131–155

19. J.N. Kuts, J.P. Gordon, S.G. Evangelides, Dynamics of dispersion managed solitons in optical communications, in *Soliton Symposium*, Kyoto (1997), pp. 183–195

Chapter 7
High-Capacity Optical Transmission Using Wavelength Division Multiplexing

Abstract The breakthrough technologies that led to the rapid increase in capacity of optical communication systems are EDFA (Erbium-Doped Fiber Amplifier), which emerged in the late 1980s, and Wavelength Division Multiplexing (WDM) technology, which became widespread in the late 1990s. The first long-distance optical amplifier repeater submarine cable system, TPC-5, used a single-channel signal of 5 Gb/s. However, as mentioned in the previous chapter, the bit rate in long-distance transmission was limited to about 40 Gb/s. Therefore, research and development for long-distance, high-capacity systems shifted to WDM systems, which can expand total capacity by increasing the number of wavelengths. In the one of the earliest WDM-based Trans-Pacific optical submarine cable systems, the China-US cable, WDM technology with a channel bit rate of 2.5 Gb/s was introduced. Compared to TPC-5's 5 Gb/s bit rate, the WDM system had to reduce the channel bit rate. This was due to the difficulty of transmitting high-speed WDM signals over long distances, caused by the fiber nonlinearity. High-speed optical signals are more severely affected by FWM (Four-Wave Mixing) and XPM (Cross-Phase Modulation) between WDM signals. Moreover, as discussed in Chap. 5, NRZ-OOK signals are not the optimal signal format for nonlinear optical fiber transmission. This chapter describes the basic configuration of WDM systems and various nonlinear suppression techniques developed to increase the channel bit rate from 10 to 40 Gb/s. The commercial WDM-based long-distance optical submarine cable system technologies that incorporate these techniques are also covered.

Keywords Wavelength division multiplexing · Gain equalization · Non-Zero Dispersion Shifted Fiber · Large-core fiber · Effective area-enlarged positive dispersion fiber · Slope compensating dispersion compensating fiber · Dispersion managed fiber

M. Suzuki and I. Morita, *Long-Haul Optical Fiber Communication Systems*, Textbooks in Telecommunication Engineering, https://doi.org/10.1007/978-3-032-19472-5_7

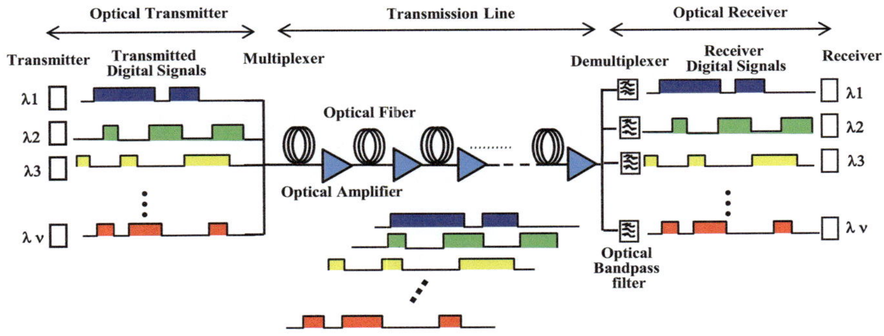

Fig. 7.1 Basic configuration of WDM system

7.1 Outline of Wavelength-Division-Multiplexing Transmission Systems

7.1.1 Basic Configuration

Figure 7.1 illustrates an overview of a wavelength division multiplexing (WDM) system. In a WDM system, multiple independent wavelength signals are combined (multiplexed) on the transmitting side, and the resulting WDM optical signal is launched into an optical transmission path composed of optical fibers and optical amplifiers. On the receiving side, the wavelength channels are separated (demultiplexed), and each wavelength signal is demodulated. The ability to share a single optical fiber and EDFA among multiple channels in the WDM signal has significantly contributed to a dramatic increase in the transmission capacity per fiber and a reduction in system costs. For achieving high-speed and large-capacity long-distance submarine cable systems, it is necessary to address issues such as mitigating the fiber nonlinearity over long distances. This involves implementing low-nonlinearity and wide-bandwidth optical fibers, low-noise and wide-bandwidth optical amplifiers, and optimizing modulation formats.

7.1.2 Wide-Band Optical Amplifiers

The gain bandwidth of an EDFA exceeds 30 nm. However, as discussed in Chap. 5, when multiple EDFAs are cascaded, the gain bandwidth becomes narrower due to the self-filtering effect. For instance, in the 9000 km TPC-5 system, over 200 EDFAs are cascaded, resulting in a gain bandwidth of just 2–3 nm. To expand the gain bandwidth of optical amplifiers for WDM systems, gain equalizers must be used to flatten the gain by compensating for the wavelenght dependence of amplifier's gain as shown in Fig. 7.2. Gain equalization can be achieved by Fourier-transforming

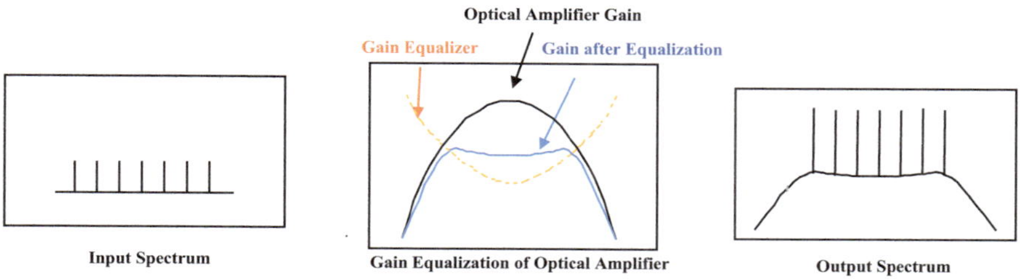

Fig. 7.2 Wider band optical amplifier using gain equalization filter

the inverse gain characteristics of the optical amplifier and synthesizing multiple periodic transfer functions corresponding to the Fourier frequency components. Gain equalizers can be implemented using technologies such as short-period and long-period fiber gratings, multi-stage Fabry–Perot optical filters, dielectric multi-layer optical filters, and planar light-wave circuit (PLC) optical filters. Key challenges in gain equalization for multi-stage optical amplifiers include reducing fitting errors, minimizing the loss of gain equalizers, and achieving low ripple characteristics.

7.1.3 Dispersion Management and Wide-Band Optical Fiber

In the 5 Gb/s single-channel TPC-5 system [1], dispersion-shifted fiber (DSF) with nearly zero dispersion at the signal wavelength was used. However, when arranging fibers with varying characteristics randomly to achieve average zero dispersion, performance degradation occurred due to issues such as excessive spectral broadening caused by self-phase modulation (SPM) and four-wave mixing (FWM) between the signal and noise. To address this, DSF with slightly negative dispersion ($D = -0.2$ ps/nm/km) was selected. A “shallow sawtooth” dispersion map was adopted, in which standard single-mode fiber (SMF) was inserted approximately every 500 km to compensate for the accumulated negative dispersion as shown in Fig. 7.3a. This approach suppressed excessive spectral broadening and nonlinear noise such as FWM. Dispersion management for DSF is generally unnecessary for terrestrial systems, making it a unique approach for long-distance submarine optical cable systems.

For WDM systems, DSF’s nonlinear suppression effects were insufficient, so non-zero dispersion-shifted fiber (NZ-DSF) with larger chromatic dispersion was introduced. In the early 2.5 Gb/s-based WDM long-distance submarine optical cable systems, NZ-DSF with dispersion of $D = -2$ ps/nm/km was employed. A “deep sawtooth” dispersion map was adopted, using SMF to compensate for accumulated chromatic dispersion (see Fig. 7.3b). As shown in Fig. 7.3b, near the center wavelength, almost ideal dispersion management can be achieved. However, for wavelengths farther from the center, the dispersion slope of the optical fiber causes the accumulated chromatic dispersion to increase with distance, as SMF

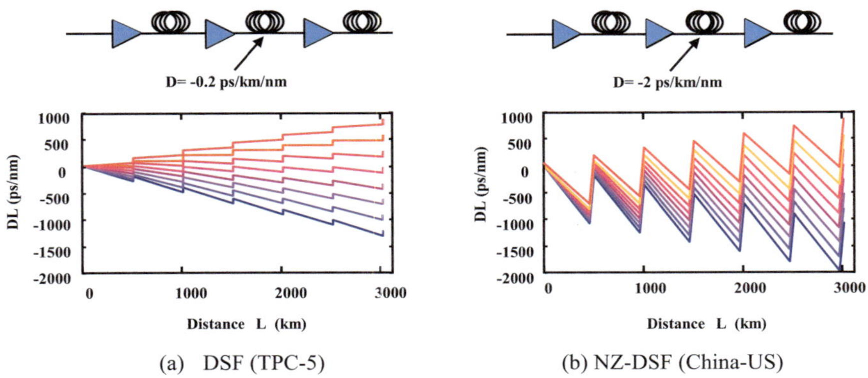

Fig. 7.3 Dispersion map of optical submarine cable systems with DSF and NZ-DSF

cannot fully compensate for the dispersion. Waveform degradation caused by the interaction between large accumulated dispersion and fiber nonlinearity cannot be compensated with linear dispersion compensation. In long-distance WDM systems, minimizing wavelength dependency in transmission characteristics is critical. This requires reducing the dispersion slope of optical fibers as much as possible when the operational bandwidth of the fibers is broadened.

7.1.4 Design Guidelines of 10 Gb/s-Based WDM Systems

The first WDM submarine optical cable system, the China-US cable [2], achieved a total capacity of 20 Gb/s (8 × 2.5 Gb/s NRZ signals) by utilizing NZ-DSF and optical amplifiers pumped at 1480 nm. To achieve higher capacity, both higher channel bit rate and more wavelength channels are required. However, both factors significantly increase the impacts of nonlinear effects and chromatic dispersion in optical fibers. For the design of 10 Gb/s-based WDM systems, technologies that can tolerate fiber nonlinearity are essential. Figure 7.4 illustrates the conceptual design guidelines for suppressing the impact of fiber nonlinearity across the entire system. For EDFAs, changing the pump wavelength from 1480 to 980 nm reduces noise by lowering the noise figure of the optical amplifiers. For optical fibers, increasing the core diameter mitigates the fiber nonlinearity directly and allows for higher fiber launch power. As shown in Fig. 7.4, this approach suppresses the impact of fiber nonlinearity and improves the optical SNR after transmission and enables the design of WDM transmission systems with high-speed signals.

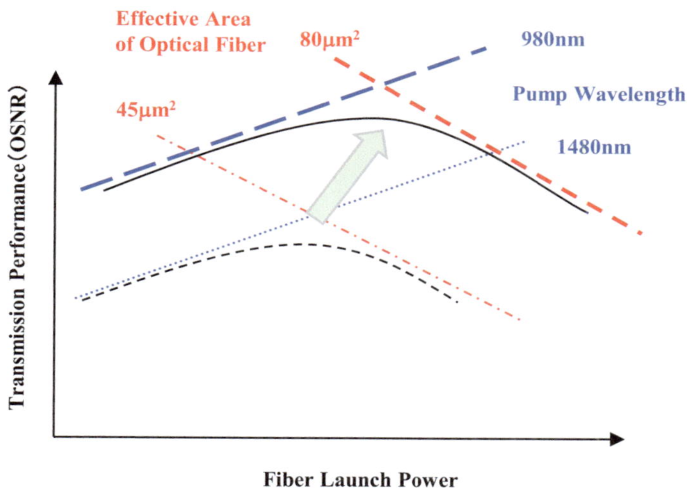

Fig. 7.4 OSNR improvement by reduction of noise generated in optical amplifiers and fiber nonlinearity

7.2 10 Gb/s, 16 WDM Transoceanic Submarine Cable Systems

In the development of WDM systems utilizing 10 Gb/s signals, it was necessary to overhaul almost all elemental technologies to suppress the degradation of transmission performance caused by the fiber nonlinearity. This included noise reduction in optical amplifiers, the development of low-nonlinearity optical fibers, and the implementation of RZ-OOK signal transmission technology. The following section describes the newly developed elemental technologies.

7.2.1 Low-Noise and Wide-Band Optical Amplifiers

When increasing the channel bit rate from 2.5 to 10 Gb/s, the required optical signal-to-noise ratio (SNR) to achieve the same bit error rate increases by 6 dB. However, simply increasing the fiber launch power by 6 dB would significantly enhance the fiber nonlinearity, leading to severe degradation of performance. To maintain the desired optical SNR while keeping the optical power level low, it was necessary to reduce the noise figure (NF) of optical amplifiers. In 2.5 Gb/s-based WDM systems, optical amplifiers pumped at 1480 nm (NF = 5.5–6.0 dB) were used. For 10 Gb/s-based WDM submarine transmission systems, the pump wavelength was changed from 1480 to 980 nm, reducing the NF to less than 4.0 dB.

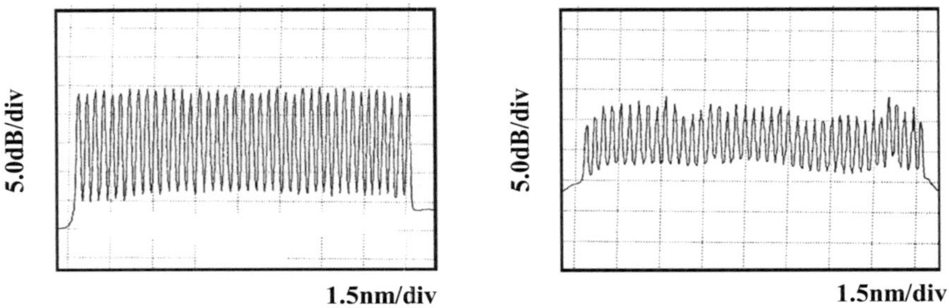

(a) 0 km (b) 18,414 km (360 repeaters)

Fig. 7.5 Optical spectrum of 40 WDM signals before after 18,414 km transmission

Furthermore, to achieve broadband optical amplifiers, gain equalization was performed using low-loss long-period fiber gratings. Figure 7.5 illustrates an example of gain equalization in an EDFA repeater system with 50 km repeater spacing. Figure 7.5a shows the optical spectrum of 40 WDM channels with a wavelength spacing of 0.3 nm. Figure 7.5b shows the optical spectrum after 18,414 km transmission (360 repeaters). Thanks to gain equalization, flat gain characteristics over approximately 12 nm were maintained even after more than 18,000 km transmission, with no significant differences observed in the optical SNR across the 40 WDM channels.

7.2.2 Low Nonlinear and Low Dispersion Slope Optical Fiber

To support high-speed and multi-wavelength transmission signals, it is necessary to expand the optical fiber wavelength band by reducing the fiber's nonlinearity and chromatic dispersion slope.

Fiber nonlinearity suppression can be achieved by increasing the effective area (A_{eff}) of the optical fiber, which reduces the light power density within the core. However, a common characteristic of optical fibers is that when A_{eff} is enlarged, the chromatic dispersion slope tends to increase as well. To address both the enlargement of A_{eff} and the suppression of the dispersion slope simultaneously, a hybrid fiber span configuration is effective, as shown in Fig. 7.6a. In this configuration, large-core fiber (LCF) with a large A_{eff} is placed immediately after the optical amplifier, where the optical power is high. After the power decreases due to the fiber loss, a Non-Zero Dispersion Shifted Fiber (NZ-DSF) with a smaller dispersion slope is placed in the latter section. This hybrid fiber span structure allows for the expansion of A_{eff} with reducing the dispersion slope. While conventional NZ-DSF has an A_{eff} of approximately 45 μm^2, the A_{eff} of LCF is expanded to about 75–80 μm^2, reducing

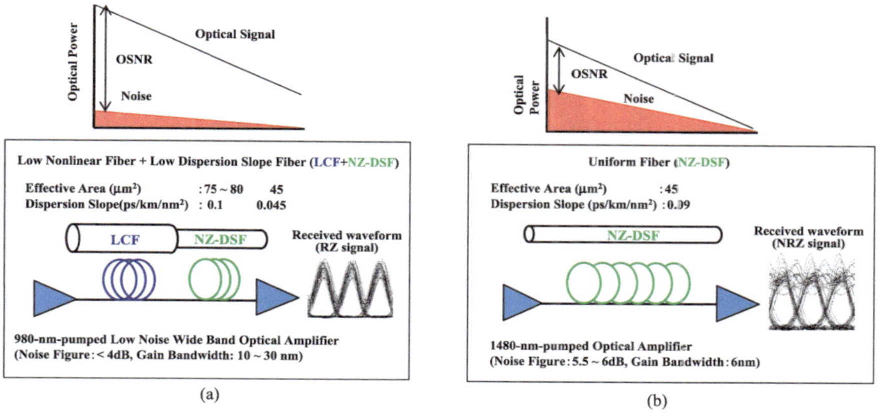

Fig. 7.6 Span configurations. **a** Hybrid span of low nonlinear and low dispersion slope fiber with low-noise optical amplifier for 10 Gb/s WDM transmission systems. **b** NZ-DSF span with conventional optical amplifier for 2.5 Gb/s WDM transmission systems

the impact of fiber nonlinearity by approximately 60%. The dispersion slope of LCF is about 0.1 ps/km/nm^2, but by setting the dispersion slope of the subsequent DSF section to a smaller value, the hybrid span configuration achieves an average dispersion slope of below 0.09 ps/km/nm^2, similar to conventional NZ-DSF. The cumulative chromatic dispersion that cannot be compensated for in the transmission path is compensated by dispersion compensation fibers evenly at both the transmitting and receiving ends.

Figure 7.6 schematically shows the optical amplifier spans of systems using (a) a low-noise optical amplifier pumped at 980 nm and a hybrid fiber span of LCF and NZ-DSF, and (b) an optical amplifier pumped at 1480 nm and standard NZ-DSF. Comparing the two, it can be seen that the low-noise amplifier and hybrid fiber span in Fig. 7.6a can raise the optical output level and suppresses the noise generated in optical amplifier, resulting in a significantly higher optical SNR compared to the traditional system in Fig. 7.6b.

7.2.3 Modulation Formats

In Chap. 6, it was mentioned that Return-to-Zero (RZ)-OOK optical signals are suitable for nonlinear transmission systems where the chromatic dispersion alternates between positive and negative values periodically. Figure 7.7 shows examples of the simulated waveforms after 7,500 km transmission of 10 Gb/s RZ-OOK and Non-Return-to-Zero (NRZ)-OOK optical signals, each with 4 wavelength multiplexing, through fiber spans shown in Fig. 7.6a with periodic dispersion compensation. In the

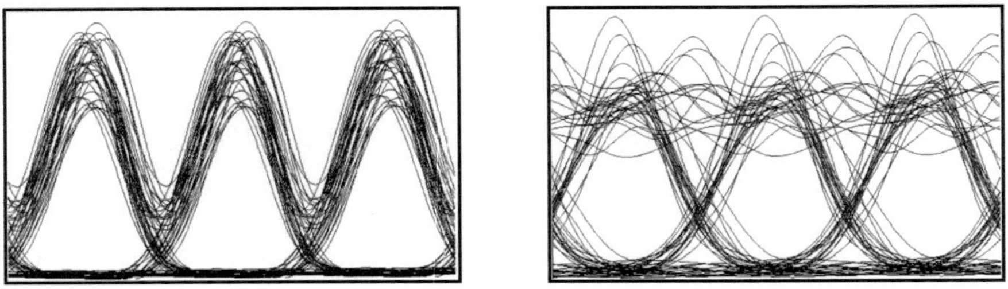

(a) RZ-OOK signals (b) NRZ-OOK signals

Fig. 7.7 Waveforms of RZ-OOK and NRZ-OOK signals after 7,500 km transmission (10 Gb/s, 4 WDM)

case of RZ-OOK optical signals, the pattern effect that causes signal degradation on the mark side is smaller compared to NRZ-OOK optical signals. Furthermore, the RZ-OOK optical signal requires approximately 1.5 dB lower optical SNR than NRZ-OOK signals to achieve the same bit error rate, indicating better noise resistance. As described in Chap. 6, the periodic stable solution for a nonlinear transmission system with periodically alternating positive and negative chromatic dispersion is the Gaussian-shaped RZ-OOK optical signal. Therefore, RZ-OOK optical signal transmission can minimize waveform degradation caused by the interaction between chromatic dispersion and fiber nonlinearity in nonlinear transmission.

7.2.4 10 Gb/s, 16 WDM Transmission Experiments

In addition to the various new optical technologies mentioned above, using an Forward Error Correction (FEC) with higher gain than the traditional Reed–Solomon code allows for increasing the channel bit rate in WDM systems from 2.5 to 10 Gb/s. Transmission experiments using a 3,600 km recirculating loop and a 9,000 km straight-line transmission line are described below.

(1) 10,850 km Transmission Using a 3,617 km Recirculating Loop

To verify the feasibility of a 10 Gb/s, 16 WDM (total capacity 160 Gb/s) system, a large-scale test-bed experiment was conducted [3] using a 3,617 km recirculating loop including 70 low-noise EDFAs pumped at 980 nm and hybrid fiber spans with an average span length of 51.7 km, as shown in Fig. 7.6a. Each hybrid span was composed of LCF (Large Core Fiber, A_{eff}: 80 μm^2, dispersion slope: 0.1 ps/km/nm^2) with average chromatic dispersion of − 2 ps/km/nm at 1550 nm and NZ-DSF (Non-zero Dispersion Shifted Fiber, A_{eff}: 45 μm^2, dispersion slope: 0.09 ps/km/nm^2) in a 1:1 ratio. Cumulative chromatic dispersion was compensated by SMFs (Single-Mode Fiber, dispersion value: + 18 ps/km/nm) placed approximately at every 500 km,

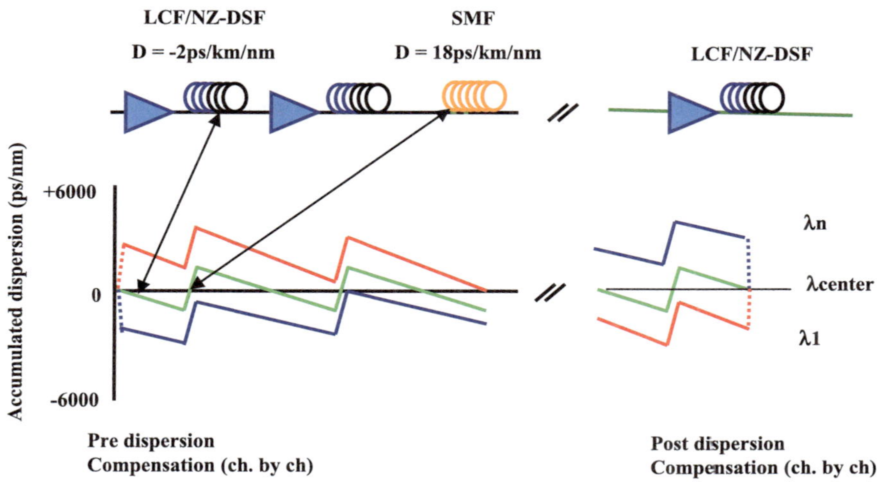

Fig. 7.8 Dispersion map for 10 Gb/s, 16 WDM transmission experiment

ensuring the system's average chromatic dispersion was zero near the center of the signal wavelength range. The dispersion map of the transmission line is shown in Fig. 7.8. At the transmission side, 16 channels of 10.7 Gb/s RZ-OOK optical signals were multiplexed with a 0.7 nm spacing to generate a 160 Gb/s WDM signal. The output power level of the optical amplifiers was set to 11 dBm. Figure 7.9 shows the transmission performance (Q values) of each channel. The average Q value for all channels was 15.5 dB, corresponding to a bit error rate (BER) of approximately 10^{-9}.

(2) 9,000 km Straight-Line Transmission Test

From Fig. 7.9, several points of improvement for system parameters aimed at commercial development can be identified. Firstly, near the central wavelength, the Q values are more than 2 dB higher compared to the wavelengths at both ends. This is likely because the transmission is closer to the propagation of dispersion-managed soliton, chirped Gaussian-type RZ-OOK pulses, which is the stable solution for nonlinear transmission systems with periodically changing chromatic dispersion. On the other hand, the cumulative chromatic dispersion of the transmission fibers at both ends of the wavelength has an absolute value of over 5,000 ps/nm, causing significant degradation of the system's performance. This indicates that waveform degradation caused by the interaction of fiber nonlinearity, chromatic dispersion, and amplifier noise during fiber transmission cannot be compensated for by linear dispersion compensation, suggesting that further suppression of cumulative dispersion is necessary.

Taking the above into consideration, a 9,000 km straight-line transmission line with several modified system parameters was constructed, and a commercialization trial was conducted. First, in the loop experiment, the latter part of the hybrid fiber

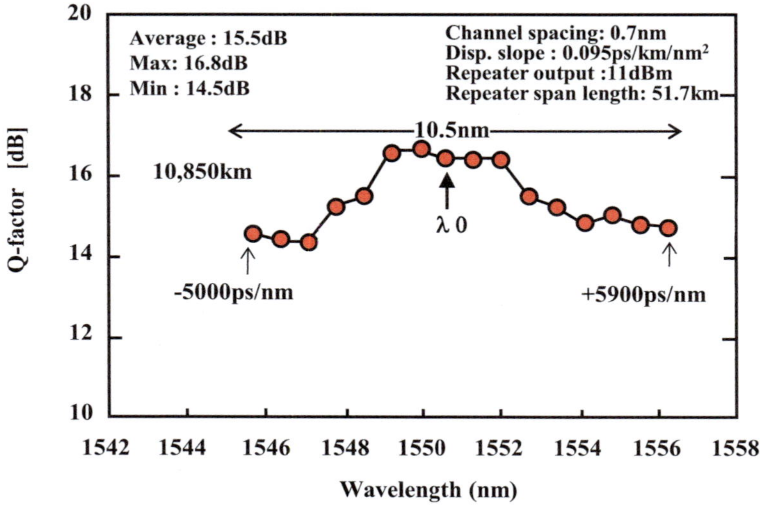

Fig. 7.9 *Q*-factors of 10 Gb/s, 16 WDM signals after 10,850 km transmission

span used conventional NZ-DSF, but this was changed to NZ-DSF with a lower dispersion slope (0.045 ps/km/nm^2), which suppressed the average dispersion slope to 0.065 ps/km/nm^2. Additionally, the wavelength spacing was changed from 0.7 to 0.6 nm, narrowing the wavelength bandwidth from 11.5 to 9 nm. As a result, the cumulative dispersion of optical transmission fibers at the edge wavelengths was suppressed to an absolute value of 3,400 ps/nm or less. To further suppress fiber nonlinearity, the fiber launch power was reduced from 11 to 9.5 dBm, and to compensate for the resulting degradation of optical SNR, the optical fiber span length was set to 45 km. The straight-line test bed, consisting of 200 EDFAs and 9,000 km of optical fiber, is shown in Fig. 7.10. This test bed was constructed in collaboration with KDD, KDD-SCS, and AT&T-SSI (Later TE-Subcom) at the time. Using this test bed, transmission performance evaluations, monitoring tests, system failure tests, long-term stability tests, and other experiments were conducted over a period of more than one year.

Figure 7.11 shows the bit error rate characteristics (*Q* value) after the transmission of 10 Gb/s, 16 WDM signals over 9,000 km. The lowest *Q* value was 16.02 dB, and it is evident that the wavelength dependence of the transmission characteristics has been significantly improved. Considering the use of FEC, this result ensured that even with margin distribution for aging degradation, manufacturing variations, and time fluctuations, a bit error rate of 10^{-13} or lower could be guaranteed after 25 years. These technologies have been adopted in the Trans-Pacific optical submarine cable systems, Japan-US and PC-1, and the Trans-Atlantic optical submarine cable systems, TAT-14, which began operation between 1999 and 2001.

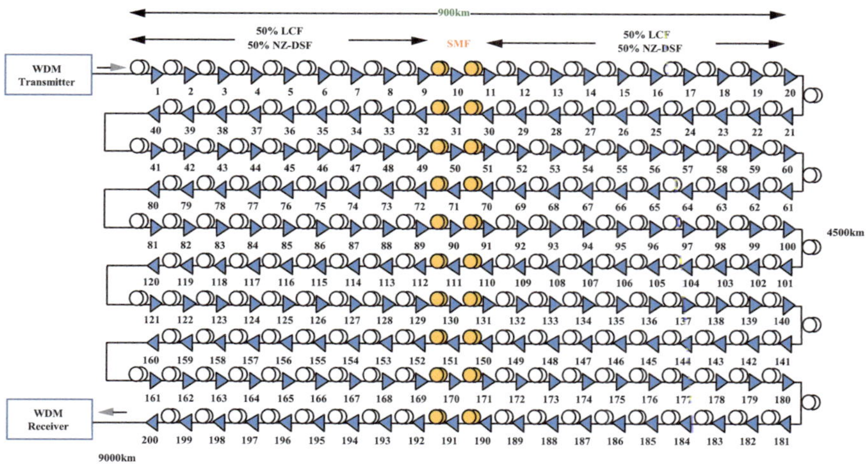

Fig. 7.10 9,000 km straight-line test bed

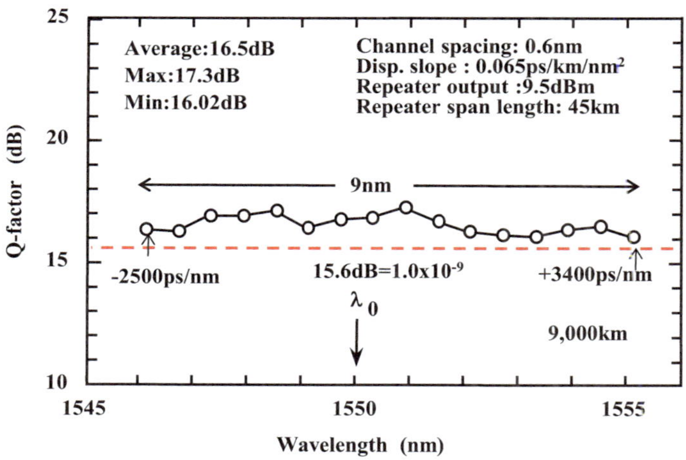

Fig. 7.11 Transmission performance in 9,000 km straight-line experiment

7.3 Terabit Transmission Using 10 Gb/s Wavelength Division Multiplexing

The performance degradation due to fiber nonlinearity and cumulative chromatic dispersion becomes more pronounced as the transmission distance increases and the signal wavelength band expands. Therefore, to transmit terabit-class large-capacity signals, in addition to further expanding the gain bandwidth of optical amplifiers and suppressing the fiber nonlinearity, new fiber dispersion management is required to reduce the dispersion slope of the optical fiber, ensuring that the chromatic dispersion remains nearly constant over a wide wavelength range.

7.3.1 Wide-Band Dispersion Management

Figure 7.12 shows the dispersion map for a terabit system. In the terabit system, each fiber span is constructed using an effective area-enlarged positive dispersion fiber (EE-SMF) and a slope compensating dispersion compensating fiber (SC-DCF) with a negative dispersion and a negative dispersion slope, as shown in Fig. 7.13. The hybrid span of EE-SMF and SC-DCF creates a dispersion-flat optical transmission line, where the dispersion remains constant across a wide signal wavelength range as shown in Fig. 7.14, enabling uniform transmission performance across a broad bandwidth.

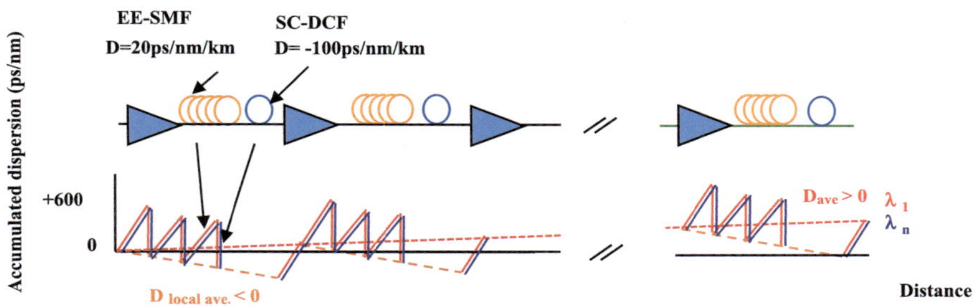

Fig. 7.12 Dispersion map of terabit WDM systems

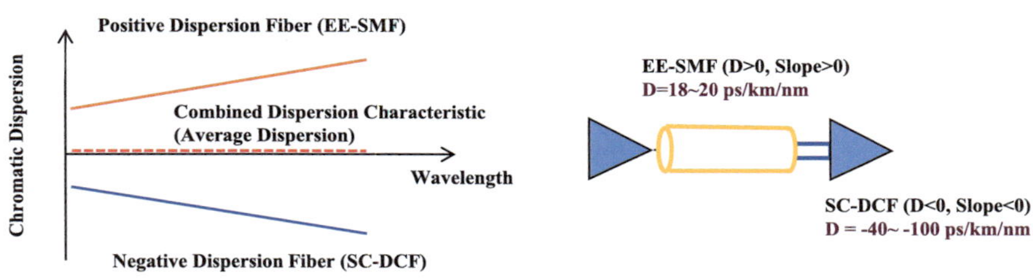

(a) Chromatic dispersion characteristics (b) Span Configuration

Fig. 7.13 Hybrid span with EE-SMF and SC-DCF

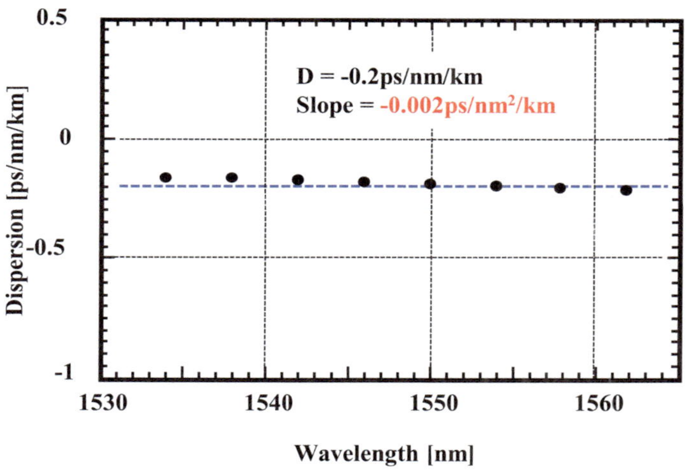

Fig. 7.14 Chromatic dispersion of hybrid span with EE-SMF and SC-DCF

7.3.2 10 Gb/s, 100 WDM, 7,750 km Transmission Experiment

A terabit loop transmission experiment was conducted using C-band optical amplifiers with the technology mentioned above [4]. At the transmitter, 100 channels of 10.7 Gb/s RZ-OOK optical signals were multiplexed at 0.3 nm wavelength intervals, with even and odd channels having orthogonal polarization. The 287 km loop transmission line consisted of seven 40 km spans of transmission fiber and nine C-band optical amplifiers, which were constructed using 980 nm-pumped EDFAs. To achieve a flat gain over a 30 nm wide band, each optical amplifier was equipped with long-period fiber gratings for gain equalization. The noise figure of the amplifiers was approximately 4.2 dB, and the output optical power was set to about + 16.5 dBm. The transmission optical fiber spans were configured as dispersion-flat hybrid fiber spans, using the two types of fibers mentioned above. Immediately after the optical amplifiers, 34 km of positive dispersion fiber, EE-SMF, with an effective area (A_{eff}) of 110 μm^2 was placed to suppress the effect of fiber nonlinearity. Following this, a dispersion slope compensating fiber (SC-DCF) of 6 km, with negative dispersion and negative dispersion slope, was used to flatten the chromatic dispersion across a broad wavelength range. As a result, the average dispersion slope at the central wavelength was reduced to less than 1/10 of that of a normal fiber, at + 0.008 ps/nm^2/km, achieving a nearly dispersion-flat optical transmission line with uniform chromatic dispersion over a 30 nm wavelength range. The average dispersion of each 40 km span was set to negative value of – 0.2 ps/nm/km as shown in Fig. 7.14, and accumulated negative dispersion in the loop transmission line was compensated by SMF so as to get slightly positive average dispersion to support dispersion-managed soliton as shown in Fig. 6.7. With this two-step dispersion management, dispersion-managed fiber equivalent to that shown Fig. 6.18 for dispersion-managed soliton transmission over NZ-DSF with negative dispersion was realized.

Figure 7.15a shows the optical spectrum after 7,750 km of transmission. Even after 243 repeaters, the bandwidth of approximately 30 nm was maintained. The average SNR across all wavelength channels was about 14.8 dB (wavelength resolution: 0.1 nm). Figure 7.15b shows the *Q*-values for each wavelength channel. The average *Q*-value across all channels was 14.3 dB without error correction. By using a transmission line with constant chromatic dispersion across the entire transmission band, nearly uniform transmission performance was achieved across all wavelength channels. Figure 7.16 shows the transmission distance dependence of *Q*-values for the center channel. This experiment confirmed the feasibility of a transoceanic terabit optical submarine cable system using only the C-band, without increasing the number of repeater components or power consumption.

The transmission technology using RZ-OOK optical signals and wide-band dispersion management techniques was adopted in the early 10 Gb/s DWDM

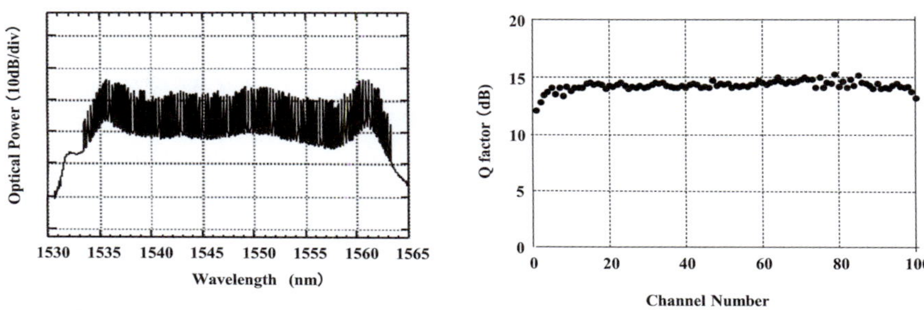

(a) Optical Spectrum of received signal (b) Q-factors after transmission

Fig. 7.15 Experiment results in 10 Gb/s, 100 WDM, 7,750 km transmission

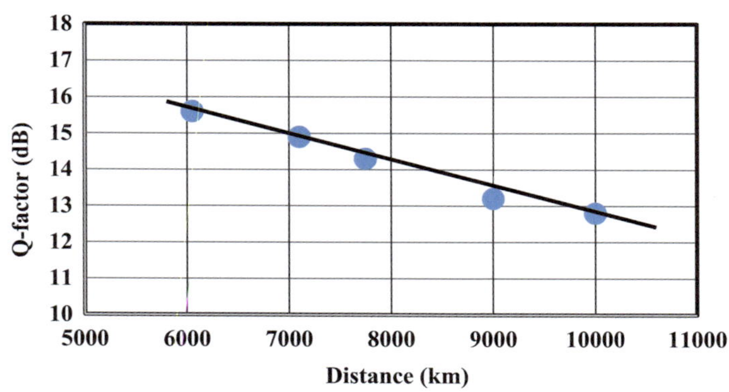

Fig. 7.16 Transmission distance dependence of *Q* factor

systems, such as the TGN-P [5]. Later, the RZ-DPSK (Differential Phase Shift Keying) optical signal, which requires less optical SNR by 3 dB compared to the conventional RZ-OOK and is capable of suppressing the fiber nonlinearity, emerged. RZ-DPSK optical signals were adopted in systems such as the Trans-Pacific cable system, UNITY, which started operation in 2010 with 96 WDM channels of 10 Gb/s.

7.4 Multi-terabit Transmission Using 40 Gb/s Wavelength Division Multiplexing

To increase the channel bit rate from 10 to 40 Gb/s, a further 6 dB improvement in optical SNR is required. The use of RZ-DPSK allows for a reduction of 3 dB in the required optical SNR compared to RZ-OOK, so it is possible to meet this requirement with the combination of high-gain Forward Error Correction (FEC) and other techniques that can reduce the required optical SNR. However, to increase capacity within the limited C-band wavelength range, improvements in spectral efficiency are needed. In this context, the following describes a 40 Gb/s, 64 WDM transmission experiment aimed at high-speed and large-capacity transmission by improving spectral efficiency. The total capacity was 2.56 Tb/s, and the transmission distance was 8,200 km [6].

The optical transmitter used 64 channels of 42.7 Gb/s CS (Carrier Suppressed)-RZ DPSK signals with 67% duty ratio and 50 GHz spacing (transmission bandwidth: 1539.8–1565.1 nm). These signals were bandwidth-limited by a periodic optical filter with a 3 dB bandwidth of 45 GHz, and the even and odd channels were combined with orthogonal polarization. Figure 7.17 shows the optical spectrum of a single-channel extracted from the WDM signal after filtering. The crosstalk in the figure comes from adjacent channels, but since the polarizations are orthogonal, the impact on the transmission performance is minimal.

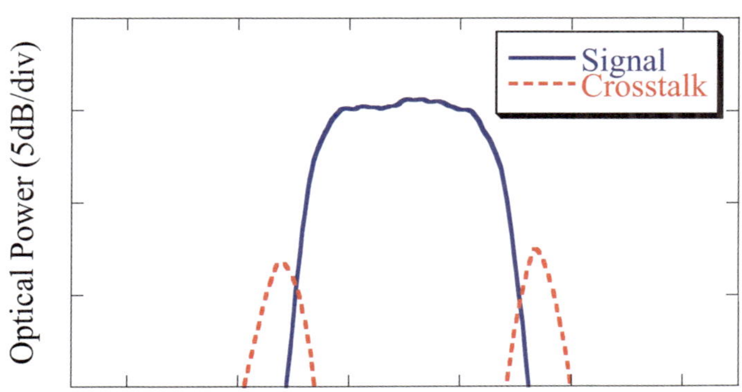

Fig. 7.17 Optical spectrum of pre-filtered CS-RZ DPSK signal

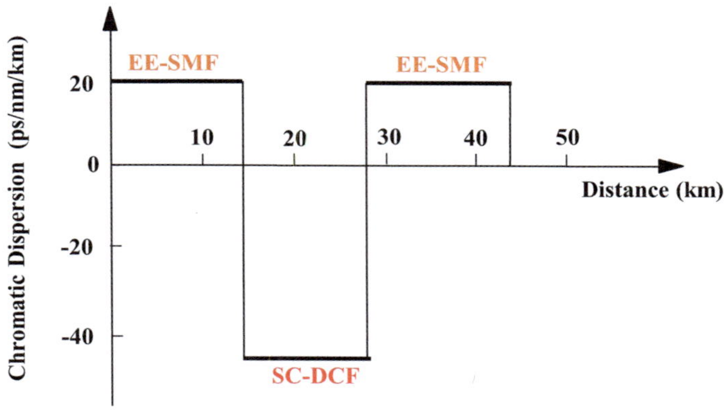

Fig. 7.18 Symmetric dispersion-managed fiber span

The 357 km long recirculating loop was mainly composed of 10 EDFA repeaters and 8 spans of 43 km symmetric dispersion-managed fiber spans. As shown in Fig. 7.18, the symmetric dispersion-managed fiber spans consisted of 14 km of EE-SMF with $A_{\mathrm{eff}} = 110\ \mu\mathrm{m}^2$, $D = 20$ ps/nm/km, and dispersion slope of 0.06 ps/km/nm^2, followed by 15 km of SC-DCF with negative dispersion and dispersion slope, with $A_{\mathrm{eff}} = 30\ \mu\mathrm{m}^2$, and another 15 km of EE-SMF. The average dispersion of the span at 1550 nm was − 0.9 ps/nm/km, and after inserting dispersion-adjustment fibers, the average dispersion of the loop was set to + 0.03 ps/nm/km. This fiber dispersion arrangement was designed to enable stable transmission of Gaussian-shaped RZ pulses, which is the steady-state solution for nonlinear transmission lines, as discussed in Chap. 6 on dispersion-managed soliton. Therefore, the deterioration of the RZ optical waveform during transmission was minimized. The output signal power of the EDFA repeaters was set to + 13 dBm.

In the optical receiver, after selecting the desired channel using an optical filter, the 42.7 Gb/s optical signal was demodulated using a 1-bit delay interferometer and received by a balanced receiver. The signal was then time-division demultiplexed using an electrical demultiplexer to separate it into four 10.7 Gb/s signals, and the average bit error rate was measured for these signals. Figure 7.19 shows the optical spectrum after 8,200 km transmission, as well as the balanced received waveform for a representative channel before and after transmission. Even after 8,200 km transmission, there was no significant signal waveform distortion, and a sufficient eye opening was observed, indicating that the system maintains excellent signal quality over long distances.

Figure 7.20 shows the transmission performance (optical SNR and Q value) after 8,200 km transmission with 64 WDM channels. A Q value of over 9.5 dB was achieved for all channels, and it was confirmed that the transmission performance exceeded the error correction threshold (BER $< 1 \times 10^{-13}$) using concatenated Reed–Solomon (RS) codes (RS(255,247) + RS(247,239)). This indicates that the system

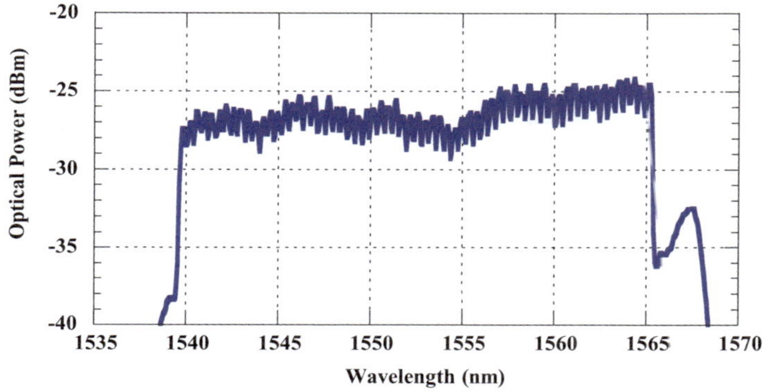

(a) Optical spectrum after 8,200 km transmission

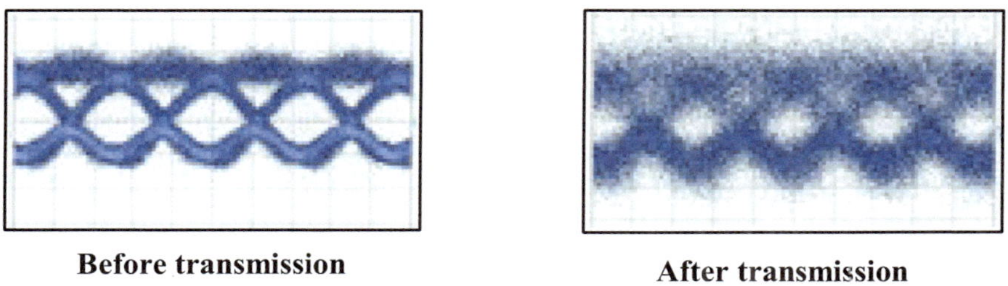

(b) Signal waveforms received by balanced receiver before and after transmission

Fig. 7.19 Optical spectrum and signal waveforms in 8,200 km transmission experiment

is capable of maintaining high-quality transmission even over long distances with advanced FEC.

7.5 Commercial Trans-Pacific Submarine Cable Systems

In Table 7.1, a comparison is made between the various technical elements of the 10 Gb/s, 16 WDM Trans-Pacific submarine cable system (such as Japan-US) and the initial 2.5 Gb/s, 8 WDM Trans-Pacific submarine cable system (such as China-US). 2.5 Gb/s-based WDM systems used NRZ-OOK signals. However, in order to improve the nonlinear tolerance, RZ-OOK signals were employed for the 10 Gb/s WDM systems. As bit rate increased, it became necessary to mitigate the fiber nonlinearity furthermore. To address this, new technologies were adopted, such as large-core fibers (LCF) and low-noise, wide-band optical amplifiers pumped at 980 nm, leading to a dramatic increase in fiber capacity, reaching up to 160 Gb/s. Furthermore, the 10 Gb/s-based WDM technology was not only deployed in Trans-Pacific and Trans-Atlantic

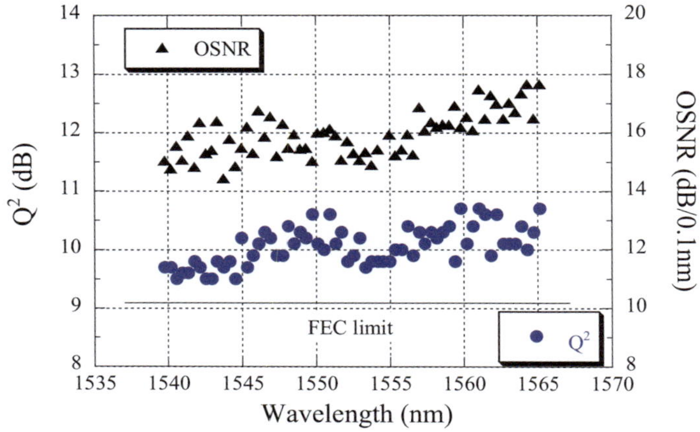

Fig. 7.20 Transmission performance after 8,200 km transmission

submarine cable systems (such as PC-1, Japan-US, TAT-14) but also commercialized in regional cable systems in the Asia region like the EAC and C2C systems, extending the number of wavelengths to 64–96 for systems spanning 3,000–4,000 km.

In Table 7.2, the key characteristics of the Trans-Pacific submarine cable systems using optical amplifiers are shown. After the 10 Gb/s, 16 WDM system discussed in Sect. 7.2, the advances in fiber dispersion management, as mentioned in Sect. 7.3, enabled the flattening of chromatic dispersion. This made it possible to achieve uniform transmission performance across a wide wavelength band of over 30 nm. This wide-band dispersion management technology was adopted in systems like TGN-P and UNITY, leading to the realization of terabit-class Trans-Pacific systems. The 40 Gb/s RZ-WDM transmission technology, introduced before the 100 Gb/s WDM technology of FASTER (which will be discussed in the next chapter using digital coherent technology), was not directly commercialized for Trans-Pacific submarine cable systems. However, it was implemented in short-distance optical submarine cable systems.

It is important to note that Table 7.2 represents the initially designed total capacity. While the underwater section of a submarine cable system cannot be modified, the terrestrial terminal equipment is often updated with technological advancements, which increases the system's capacity. For example, the Japan-US cable system initially operated with 16 WDM channels of 10 Gb/s RZ-OOK signals, but after the emergence of RZ-DPSK, the system was upgraded to expand to 32 WDM channels through replacement of the terminal equipment.

Figure 7.21 shows the transition of the fiber-pair capacity of Trans-Pacific submarine cable systems from 1980s to 2020s. The impact of the introduction of the 10 Gb/s WDM technologies is clearly visible, with a steep increase in capacity in the early 2000s. The 10 Gb/s-based WDM system, used in systems from PC-1 to UNITY, relies on dispersion management with periodic positive and negative chromatic dispersion, allowing the use of RZ-OOK optical pulses. Over the 17 years from PC-1 in 1999 to FASTER in 2016, dispersion management considering fiber nonlinearity and the steady-state solution of Gaussian-shaped RZ pulses were used. This indicates that the dispersion-managed soliton transmission technology introduced in Chap. 6 forms the foundation for these systems.

Finally, Fig. 7.22 illustrates the current global submarine cable network. The global optical submarine cable network has provided a high quality, wide-bandwidth communication environment capable of accommodating increasing traffic during the period when the internet rapidly expanded. Today, the global submarine cable network is an essential backbone infrastructure that supports our modern information society.

Table 7.1 Main features of commercialized 2.5 and 10 Gb/s WDM systems

	2.5G WDM	10G WDM
System configuration	2.5 Gb/s, 8 WDM (~ 10,000 km)	10 Gb/s, 16 WDM (7 ~ 9,000 km) 10 Gb/s, 64–96 WDM (3 ~ 4000 km)
Applied system	China-US, SEA-ME-WE 3	TAT-14, Japan-US, PC-1 (160 Gb/s) EAC (640 Gb/s), C2C (960 Gb/s)
Optical repeater	Pump wavelength: 1480 nm Noise figure: 6 dB Bandwidth: 6 nm → High speed transmission is difficult due to low nonlinear tolerance and low OSNR	Pump wavelength: 980 nm Noise figure: < 4 dB Bandwidth: > 10 nm → High speed transmission is possible due to high nonlinear tolerance and high OSNR
Optical cable	Span configuration: Single type fiber Effective area: ~ 45 μm^2 Dispersion slope: ~ 0.09 ps/km/nm^2 → Higher speed transmission is difficult due to large nonlinear distortion	Span configuration: Hybrid span with large A_{eff} fiber and low dispersion slope fiber Effective area: 75–80 μm^2, 45 μm^2 Dispersion slope: ~ 0.09 ps/km/nm^2, ~ 0.045 ps/km/nm^2 → Higher speed transmission is possible by introducing low nonlinear and low dispersion slope fibers
Transmitter and receiver	2.5 Gb/s NRZ signal FEC gain: ~ 6 dB → Large pattern effect due to fiber nonlinearity	10 Gb/s chirped Gaussian RZ signal FEC gain: ~ 8 dB → Small pattern effect due to fiber nonlinearity

Table 7.2 Key characteristics of Trans-Pacific optical submarine cable systems

System	In year	Capacity/ F.P. (bps)	Bit rate/λ (bps)	No. of λ	No. of F.P.	Cable capacity (bps)	Modulation format	Fiber type
TPC-5	1996	5G	5G	1	2	10G	NRZ-OOK	DSF (− D/ + D)
China-US	2001	20G	2.5G	8	4	80G	NRZ-OOK	NZ-DSF (− D/+ D)
PC1	1999	160G	10G	16	4	640G	RZ-OOK	NZ-DSF (− D/+ D)
Japan-US	2001	160G	10G	16	4	640G	RZ-OOK	NZ-DSF (− D/+ D)
TGN-P	2002	640G	10G	64	8	5.12T	RZ-OOK	DMF (+ D/ − D)
Unity	2010	960G	10G	96	5	4.8T	RZ-OOK (DPSK)	DMF (+ D/ − D)
Faster	2016	10T	100G	100	6	60T	DP-QPSK	SMF (+ D)

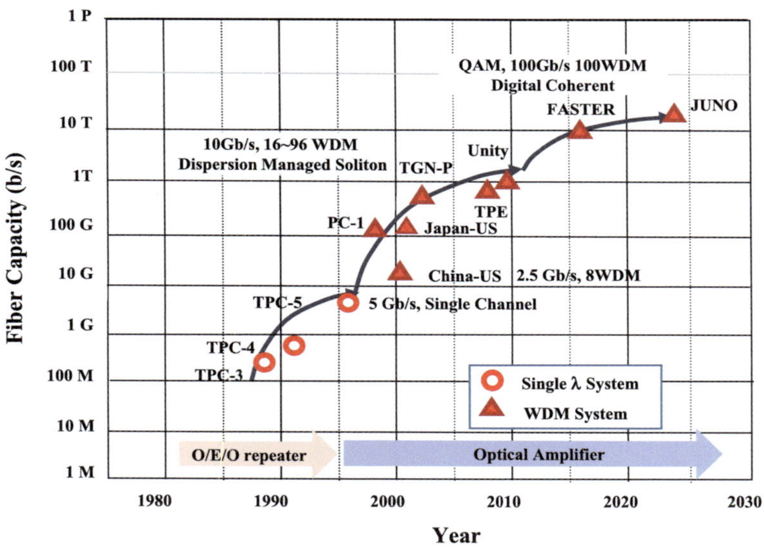

Fig. 7.21 Evolution of fiber capacity in Trans-Pacific optical submarine cable systems

Problems

7.1 What total output power should deliver an optical amplifier to ensure 0.5 dBm power per channel in the presence of 20 channels in WDM transmission?

7.2 Explain the reason why the dispersion management is applied in WDM transmission systems.

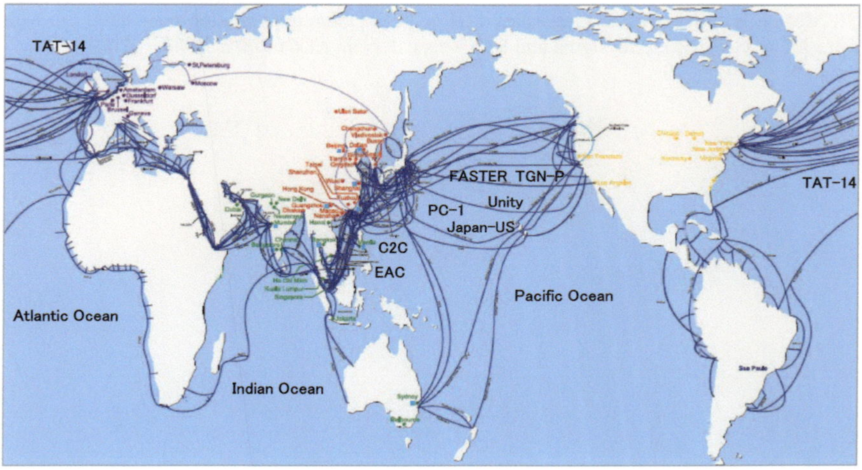

Fig. 7.22 Global optical submarine cable network

7.3 Draw the dispersion map such as Fig. 7.3 for WDM transmission with 1-nm spaced 11 channels using positive and negative dispersion fiber with the following parameters.

- Chromatic dispersion and dispersion slope of the negative dispersion fiber: − 2 ps/nm/km and 0.1 ps/nm^2/km.
- Chromatic dispersion and dispersion slope of the positive dispersion fiber: 20 ps/nm/km and 0.1 ps/nm^2/km.
- Dispersion compensation interval: 100 km.
- Average chromatic dispersion at center channel: 0 ps/nm/km.

References

1. S. Akiba, A long haul optical amplifier system engineering based on EDFA technology, in *OAA'93*, Yokohama (1993), pp. 110–113
2. M. Suzuki, H. Kidorf, N. Edagawa, H. Taga, N. Takeda, I. Morita, S. Yamamoto, E. Shibano, T. Miyakawa, E. Nazuka, M. Ma, F. Kerfoot, R. Maybach, H. Adelmann, V. Arya, C. Chen, S. Evangelides, D. Gray, B. Pedersen, A. Puc, 170 Gb/s transmission over 10850 km using large core transmission fiber, in *OFC'98, PD17* (1998)
3. T. Tsuritani, Y. Yamada, A. Agata, N. Takeda, N. Edagawa, M. Suzuki, 1 Tbit/s (100 × 10.7 Gbit/s) transpacific transmission over 7,750 km using single-stage 980 nm-pumped C-band optical repeaters without forward error correction, in *OECC'2000, 11A2-3* (2000), pp. 22–23
4. B. Bakhshi, M.F. Arend, M. Vaa, E.A. Golovchenko, D. Duff, H. Li, S. Jiang, W.W. Patterson, R.L. Maybach, D. Kovsh, 1 Tbit/s (101 × 10 Gbit/s) transmission over transpacific distance using 28 nm C-band EDFAs, in *OFC'2001, PD21* (2001)
5. M. Suzuki, N. Edagawa, Dispersion-managed high-capacity ultra-long-haul transmission. IEEE J. Lightwave Technol. **21**, 916–929 (2003)

6. I. Morita, N. Edagawa, 50 GHz-spaced 64 × 42.7 Gbit/s transmission over 8200 km using pre-filtered CS-RZ DPSK signal and EDFA repeaters, in *ECOC2003, Th4.3.1* (2003)

Chapter 8
High Spectral Efficiency Transmission Using Digital Coherent Detection

Abstract Coherent detection can detect both the amplitude and phase of an optical signal, offering higher receiver sensitivity. For this reason, its application to optical communication was actively explored worldwide during the 1980s. However, due to technological limitations at the time, practical implementation was difficult. With the emergence of optical amplifiers, the need for coherent detection diminished, and most research efforts were suspended. After 2000, as demand for higher-capacity optical fiber communication systems increased, the conventional direct detection method became insufficient to meet this demand. This led to renewed interest in optical coherent communication. A new approach, known as digital coherent detection, was proposed to address the challenges of traditional coherent communication through advanced digital signal processing technologies. With the introduction of digital coherent detection, it became possible not only to employ advanced modulation formats that utilize both the amplitude and phase of light, but also to compensate for all of the chromatic dispersion accumulated along the transmission path using DSP at the receiver. As a result, dispersion management in the transmission line became unnecessary, enabling the minimization of fiber loss and nonlinearity. This chapter explains digital coherent detection technologies which have become indispensable in recent high-capacity optical communication systems.

Keywords Digital coherent detection · Digital signal processing · Geometric shaping · Probabilistic shaping · Polarization multiplexing · FASTER · Open submarine cable · Gaussian or Generalized Signal-to-Noise Ratio

8.1 Outline of Digital Coherent Detection Systems

Coherent detection involves receiving the optical signal by interfering the received optical signal with a local light generated by a light source installed within the receiver. In contrast to the direct detection, which detects only the intensity (amplitude information) of the received optical signal, the coherent detection method can detect both the amplitude and phase information of the received optical signal.

M. Suzuki and I. Morita, *Long-Haul Optical Fiber Communication Systems*, Textbooks in Telecommunication Engineering, https://doi.org/10.1007/978-3-032-19472-5_8

Coherent detection is commonly used in wireless communication systems, and its high sensitivity has attracted attention for use in optical communication systems, with research on it beginning in the 1980s. However, compared to wireless communication systems, where the carrier frequency is much lower, optical communication systems have much higher carrier frequencies, making it difficult to achieve lasers with significantly small phase noise (very narrow linewidth), microwave circuits for waveform equalization, and phase synchronization between the signal light and local light. As a result, it was not easily commercialized. Later, with the emergence of optical amplifiers, the demand for high-sensitivity reception was relaxed, and research into coherent communication in optical systems slowed. From 2000 onward, as the demand for higher speed transmission from 40 to 100 Gb/s increased, the introduction of high-sensitivity phase modulation schemes became indispensable for optical communication systems, leading to renewed interest in coherent detection. At this time, digital signal processing technology also made significant progress, showing that the issues of conventional coherent detection could be addressed through high-speed digital signal processing [1, 2]. The use of digital signal processing in phase synchronization and waveform equalization, for example, led to what is now called digital coherent detection.

Figure 8.1 shows the basic configuration of a digital coherent receiver. The analog optical signals input to the receiver are first interfered with the local light using an optical hybrid and then converted into digital signals using an analog-to-digital converters. Digital signal processing is then used to estimate the phase of the carrier wave and equalize the signal. As a result, the previous problem of needing to track the phase changes of the signal light with the local light is no longer necessary. Additionally, by introducing a polarization diversity configuration (Fig. 8.2) that uses a polarization beam splitter to simultaneously receive the two orthogonal polarization components of the received optical signal, the issue of polarization dependence in the receiver can also be resolved.

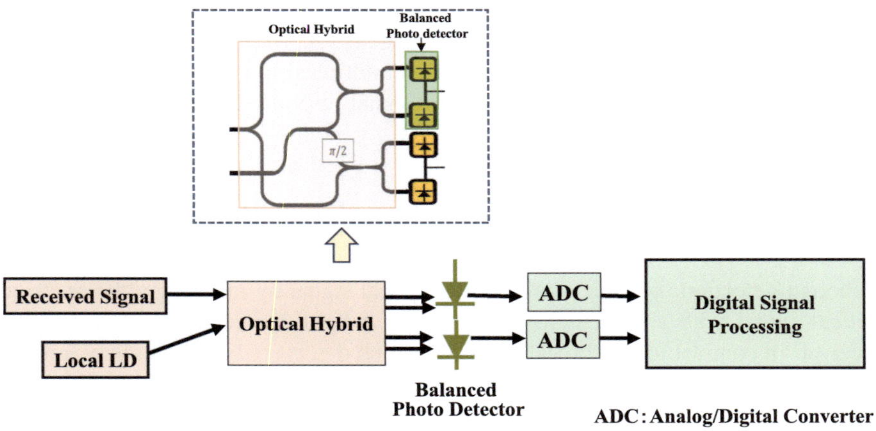

Fig. 8.1 Basic configuration of digital coherent receiver

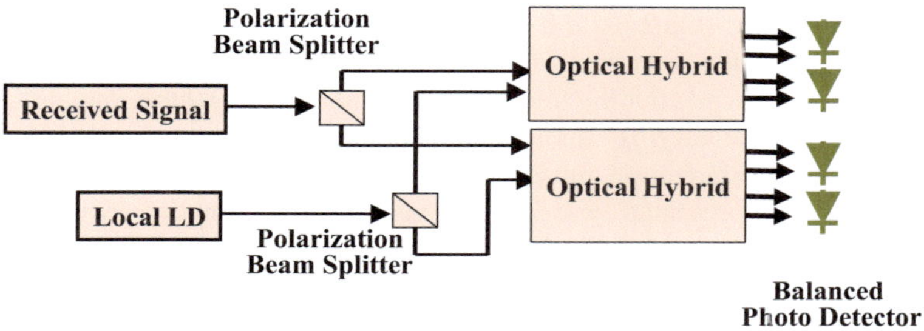

Fig. 8.2 Polarization diversity configuration

Figure 8.3 shows an example of digital signal processing in a digital coherent receiver. The optical output signals from the optical hybrid are first converted into electrical signals by photodetectors and then digitized by Analog-to-Digital Converters (ADCs). Typically, the sampling rate of the ADC is set to twice the symbol rate of the modulated signal. After digitization, signal processing is performed using the digitized signal, including chromatic dispersion compensation, polarization demultiplexing and equalization, clock recovery, and frequency/phase offset compensation, followed by signal decision.

The chromatic dispersion of the transmission line can be compensated using a Finite Impulse Response (FIR) filter, as shown in Fig. 8.4. The number of taps required for the FIR filter is proportional to the amount of dispersion compensation and the square of the symbol rate. In particular, for long-distance transmission systems such as trans-oceanic submarine cable systems, the dispersion compensation amount can exceed several thousand ps/nm, and the computational scale for dispersion compensation using an FIR filter becomes enormous. Therefore, to improve efficiency, frequency-domain filters using Fourier transforms and inverse Fourier transforms are commonly used. In this case, the coefficients of the frequency-domain

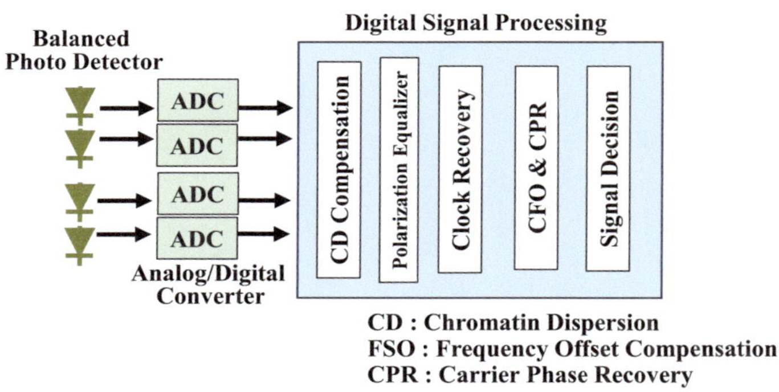

Fig. 8.3 Digital signal processing in the receiver

filter (C_k) are fixed to compensate for most of the chromatic dispersion of the transmission line, while the remaining dispersion is typically compensated by a time-domain equalizer placed after the frequency-domain filter. This structure also allows for adaptation to time-varying conditions.

The effects of polarization mode dispersion, polarization-dependent loss, and birefringence are represented by a 2×2 transfer function matrix. To compensate for these effects, an inverse matrix can be implemented using a butterfly-structured FIR filter, as shown in Fig. 8.5. Since the polarization state of the optical signal changes rapidly during fiber propagation, it is necessary to update the FIR filter coefficients to track these changes. Algorithms used for this purpose include the Constant Modulus Algorithm (CMA) and Least Mean Square (LMS) algorithm, among others. For QPSK signals with constant amplitude, CMA, which has low computational complexity and allows blind equalization, is widely used. In CMA, the tap coefficients are updated so that the absolute value of the complex amplitude remains constant. However, when applying CMA to polarization-multiplexed transmission systems, it cannot completely solve the singularity problem where the two outputs converge to the same polarization because each polarization is independently controlled. To address this issue, methods such as inserting known training signals between data signals have been proposed.

In a digital coherent receiver, since the phase of the local light is not tracked to the signal light, the output light from the optical hybrid does not have matching

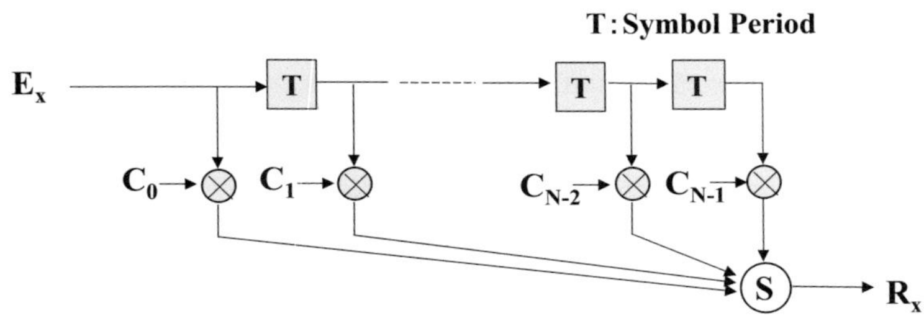

Fig. 8.4 FIR filter

Fig. 8.5 Butterfly-type FIR filter

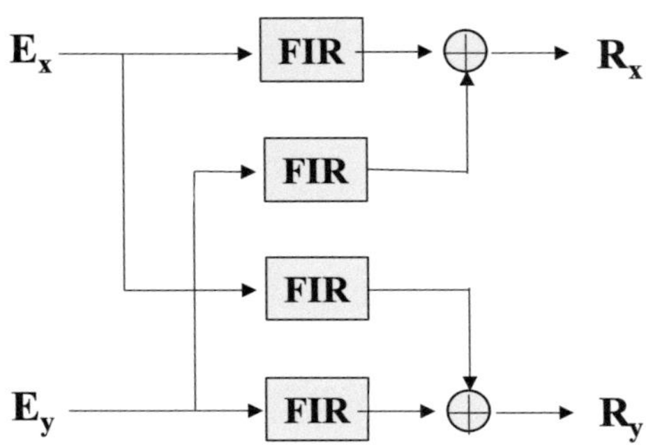

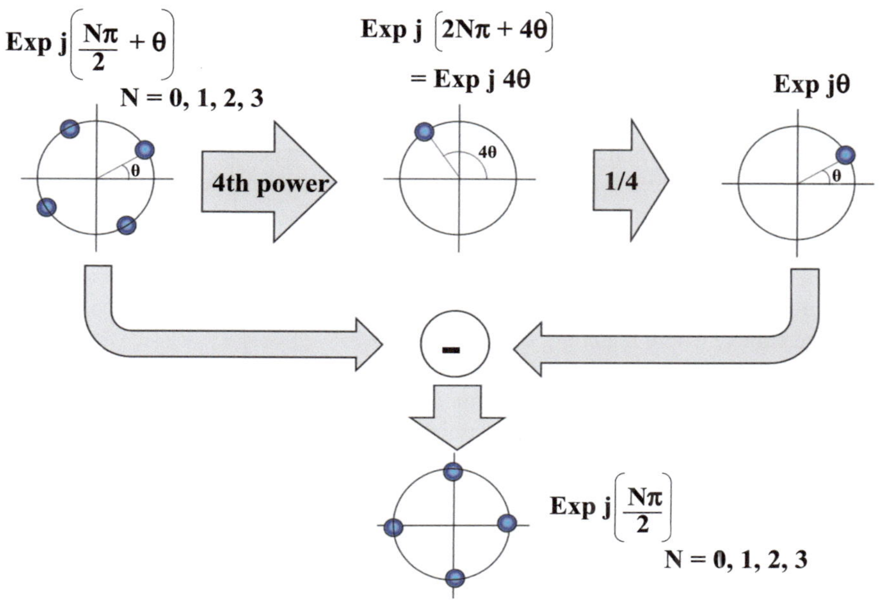

Fig. 8.6 Phase estimation of QPSK signal

frequency/phase with the local light. Therefore, for demodulation, it is necessary to detect and synchronize the frequency/phase offset between the signal and the local light. An M-phase-modulated signal can have its phase modulation removed by raising it to the Mth power. An example of a QPSK signal is shown in Fig. 8.6. If there is a phase offset of θ in a QPSK signal, raising the signal to the fourth power makes the phase modulation component an integer multiple of 2π, thus yielding 4θ. By dividing this value by 4 and subtracting it from the original signal, the offset can be compensated.

The phase estimation method using exponentiation shown in Fig. 8.6 can only be applied to phase-modulated signals, so for QAM signals, decision-directed algorithms are often used.

The digital coherent detection systems have not only high sensitivity but also the features described in Sects. 8.1.1–8.1.3.

8.1.1 Applying Various Modulation Formats

To extend the transmission distance of multi-level modulation signals such as quadrature amplitude modulation (QAM), improving the OSNR is essential. To enhance the OSNR, increasing the fiber launch power is effective. However, in long-distance optical fiber communication systems, increasing the fiber launch power also leads to larger signal degradation due to fiber nonlinearity, thus limiting the maximum fiber launch power. Therefore, along with improving the received OSNR through measures such as reducing fiber losses and minimizing noise generated in optical amplifiers, it is also important to reduce the required signal quality by enhancing the error correction performance of forward error correction codes.

One way to reduce the required OSNR for multi-level modulation signals is to optimize the signal constellation. As shown in Fig. 8.7a, in a typical QAM signal, the signal points are evenly distributed on the IQ plane. However, in the case of geometric shaping (GS), as shown in Fig. 8.7b, more signal points are placed near the center point, where the energy is low, and fewer signal points are placed farther from the center, where the energy is higher.

On the other hand, in the case of probabilistic shaping (PS), as shown in Fig. 8.7c, the signal point configuration is the same as that of a typical QAM signal. However, the probability at which each signal point is assigned is varied. The probability of assigning signal points closer to the center point, where the energy is lower, is made higher than the probability of assigning signal points farther from the center, where the energy is higher. By adopting this configuration, the overall signal energy is reduced while maintaining a constant signal point spacing (Euclidean distance). This implies that the Euclidean distance will expand when the signal energy is kept constant, leading to an improvement in signal quality.

Furthermore, as shown in Fig. 8.8, the probability at which each signal point is assigned can be freely adjusted. Therefore, by optimizing the probability distribution based on the OSNR obtained from the transmission system, the bit rate can be flexibly adjusted. This flexibility is one of the advantages of probabilistic shaping.

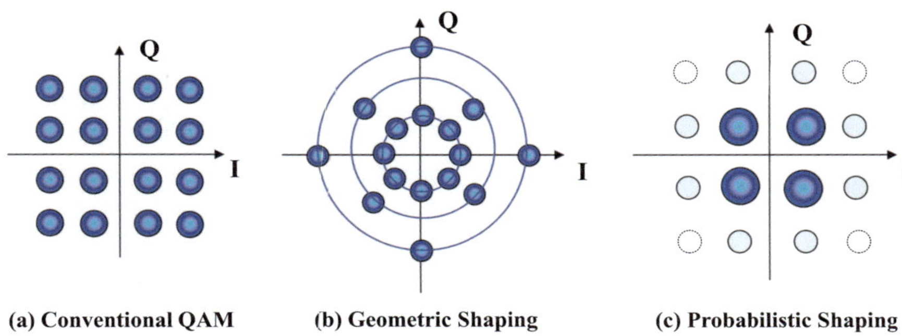

Fig. 8.7 Optimization of constellation

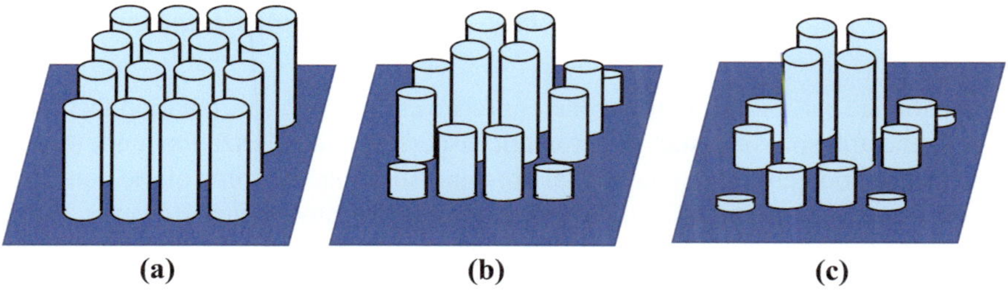

Fig. 8.8 Changes in the symbol probabilities in probabilistic shaping

8.1.2 Applying to Polarization Multiplexing Signals

In a digital coherent receiver with a polarization diversity configuration, it is possible to perform polarization demultiplexing using digital signal processing after simultaneously receiving two orthogonal polarization components. Although the polarization state of the optical signal fluctuates during fiber propagation, the orthogonality of the polarizations is maintained. As shown in Fig. 8.9, a polarization multiplexing transmission scheme can be used to transmit independent data using two orthogonal polarization components. In polarization multiplexing transmission, the transmission capacity using the same bandwidth can be doubled (doubling the spectral efficiency). Therefore, in digital coherent transmission systems, this scheme is commonly used as a method for expanding transmission capacity.

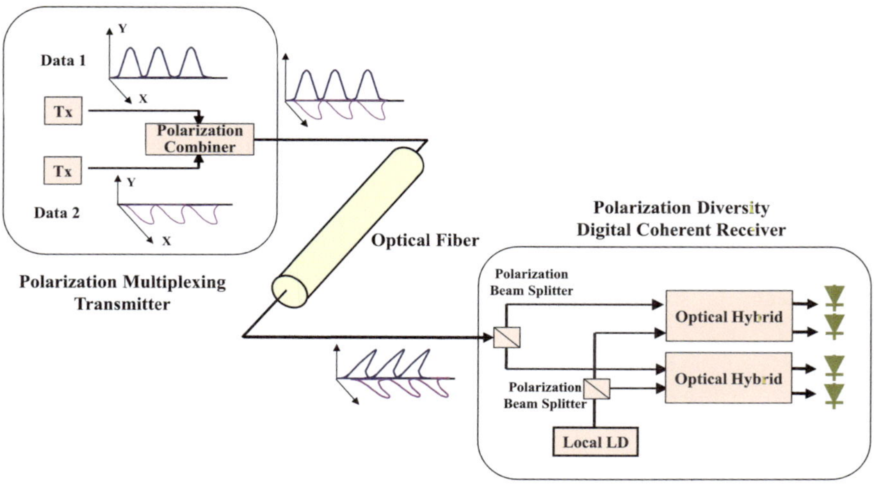

Fig. 8.9 Polarization division multiplexing system

8.1.3 *Compensation for Transmission Impairment Factors*

As mentioned earlier, in a digital coherent receiver, the linear degradation factors in optical fiber transmission, such as chromatic dispersion and polarization mode dispersion, can be compensated in the digital domain. Additionally, some of the nonlinear effects can also be compensated through digital signal processing. However, as the transmission distance increases, the computational load required for compensation grows, which becomes a challenge. In particular, compensating for fiber nonlinearity such as cross-phase modulation between wavelength-division-multiplexed signals requires receiving multiple wavelength-multiplexed signals at once and performing digital signal processing, which significantly increases the processing load, posing a challenge.

8.2 Trans-Pacific Submarine Cable System Using Digital Coherent Detection

FASTER, which began commercial service in 2016, is the first Trans-Pacific submarine cable system designed with the introduction of digital coherent detection technology. The main specifications of FASTER are shown in Table 8.1.

In FASTER, digital coherent detection technology allows for the transmission of 100 channels of 100 Gb/s signals with an initial design cable capacity of 60 Tb/s (10 Tb/s × 6 fiber pairs). The long-distance transmission of approximately 9,000 km for the 100 Gb/s signals is made possible by the technologies described below.

Table 8.1 Main specifications of FASTER

Investors	KDDI (Japan), China Mobile International (China), China Telecom Global (China), Global Transit Communications (Malaysia), Google (USA), Singtel (Singapore)
System length	About 9000 km
Designed fiber capacity	10 Tb/s (100 Gb/s × 100WDM)
Number of fiber pairs	6
Designed cable capacity	60 Tb/s
Date of service start	June 30, 2016
System supplier	NEC

8.2.1 Transmitter and Receiver Technologies

The basic configuration of the optical transmitter is shown in Fig. 8.10. In the optical transmitter, continuous wave light generated by a laser source is split into two paths for the X-polarized and Y-polarized light. Each of the split light is modulated using a vector modulator at a symbol rate of 25 Gbaud with a 4-phase-shift keying (QPSK) modulation format to generate 50 Gb/s optical modulated signals. Then, the optical modulated signals generated by the two vector modulators are combined using a polarization combiner, resulting in 100 Gb/s polarization-multiplexed QPSK signals. The actual symbol rate and bit rate are increased by adding redundant bits for forward error correction codes. In FASTER, an error correction code with higher correction capabilities than conventional codes has been introduced, reducing the OSNR required to achieve the bit error rate required by commercial systems.

The basic configuration of the optical receiver is shown in Fig. 8.11. In a digital coherent receiver with a polarization diversity configuration, digital signal processing compensates for accumulated chromatic dispersion in the transmission line, demultiplexes polarization-multiplexed signals, and compensates for polarization mode dispersion.

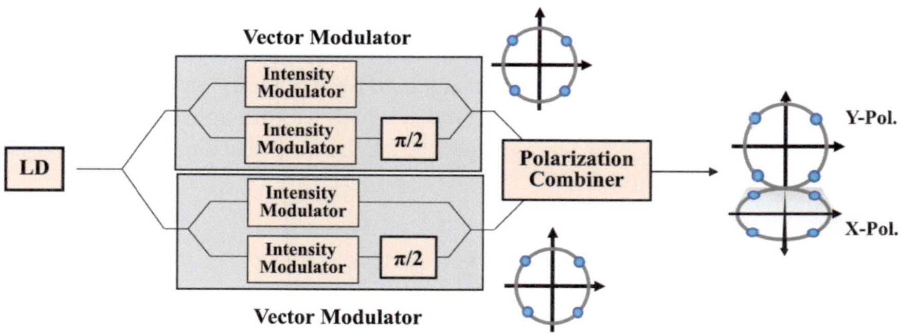

Fig. 8.10 Basic configuration of optical transmitter

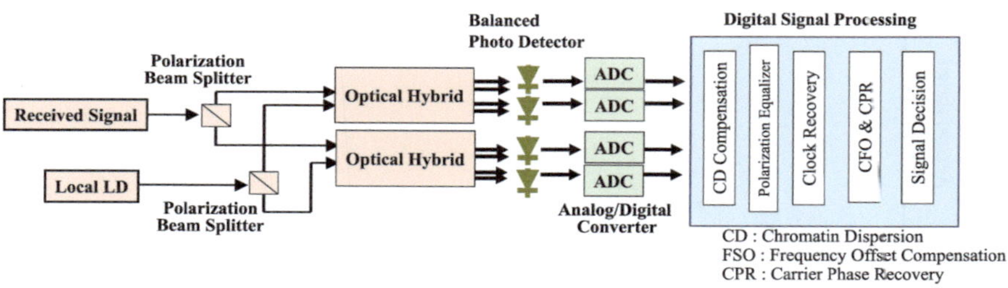

Fig. 8.11 Basic configuration of optical receiver

In the initial design of FASTER, the system was designed to multiplex up to 100 channels of 100 Gb/s signals at a 40 GHz channel spacing, achieving a spectral efficiency of 2.5 b/s/Hz.

FASTER was designed with the foresight that, while the submarine cable under sea would remain in place, the transmitter and receiver equipment could be upgraded to the latest technology to expand the capacity. In fact, it has been reported that when the modulation format was changed from QPSK to 8QAM and the channel spacing between WDM signals was narrowed to 37.5 GHz, the spectral efficiency was improved to 4 b/s/Hz [3].

Additionally, although based on offline processing, transmission experiments using 64QAM signals with probabilistic shaping (discussed in the previous section) have been conducted. In this case, the spectral efficiency was further improved to 6 b/s/Hz [4].

8.2.2 Transmission Line Technologies

The transmission line configuration of FASTER is shown in Fig. 8.12. The transmission line consists of positive dispersion fibers with low loss and low fiber nonlinearity along with C-band EDFA repeaters. The repeater spacing is approximately 60 km. The repeaters used in FASTER are equivalent to those used in earlier generations of optical submarine cable systems, such as UNITY. On the other hand, regarding the transmission fiber, dispersion management, which was previously used in WDM submarine cable systems, is not employed. Instead, the transmission line is composed of a single type of optical fiber.

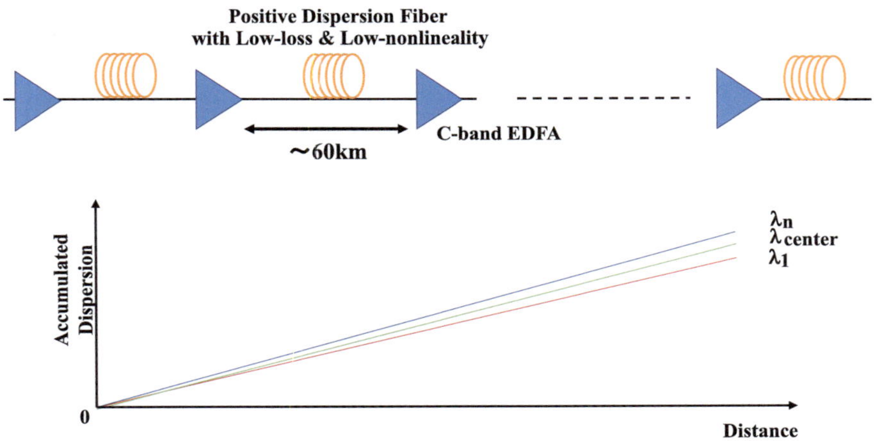

Fig. 8.12 Transmission line configuration and dispersion map

The chromatic dispersion of the positive dispersion fiber is approximately 20 ps/nm/km, and the accumulated chromatic dispersion over the Trans-Pacific distance reaches nearly 200,000 ps/nm. This accumulated chromatic dispersion is fully compensated by digital signal processing within the digital coherent receiver.

Using only one type of optical fiber for the transmission line can not only simplify the transmission line configuration but also reduce loss and fiber nonlinearity in the transmission line.

In a transmission line that uses dispersion-managed fibers consisting of positive dispersion fiber and dispersion compensation fiber (DCF), the loss of the DCF is greater compared to the positive dispersion fiber, leading to an inevitable increase in overall transmission loss. By compensating for the chromatic dispersion with digital signal processing, it is possible to eliminate the DCFs, thus avoiding the loss associated with them.

Additionally, the effective area of optical fibers differs between positive dispersion fibers and DCF. The impact of fiber nonlinearity in DCF with smaller effective areas cannot be ignored. However, by eliminating the DCFs, this impact can also be avoided. Furthermore, by not performing dispersion compensation in the transmission line, the accumulated chromatic dispersion in the transmission line never becomes close to zero, reducing the occurrence of phase matching, which in turn diminishes the impact of nonlinear effects during transmission.

In addition to the above reasons, by removing the constraint of chromatic dispersion in the design of optical fibers, it is possible to use positive dispersion fibers optimized for low loss and low nonlinearity. In FASTER, pure silica core fibers with an expanded effective area of 130 μm^2 are used for transmission. In contrast, standard single-mode fiber (SMF) with Ge doped to the core for refractive index control typically has an effective area of around 80 μm^2. By eliminating dopant from the core and expanding the effective area, the fiber is optimized for lower loss and reduced fiber nonlinearity.

8.3 Advanced Technologies for Higher-Capacity Submarine Cable Systems

In FASTER, the transmitter and receiver equipment are upgraded to achieve higher capacity. However, in order to further increase the capacity of trans-oceanic submarine cable systems, improvements in the transmission line are also being pursued.

In FASTER and previous trans-oceanic submarine cable systems, C-band Erbium-Doped Fiber Amplifier (EDFA) repeaters have been used. To expand the bandwidth available for the system, the introduction of C + L-band EDFA systems, which combine C-band EDFA with L-band EDFA (operating at longer wavelengths than C-band), is being advanced.

Figure 8.13 shows the configuration of a C + L-band EDFA. WDM signals ranging from the C-band to the L-band are first separated into the C-band and L-band using a band demultiplexer. Then, each band is amplified by an EDFA designed for that specific band. After amplification, the signals are combined using a band multiplexer and transmitted through the transmission line. The L-band EDFA is realized by using 5–10 times longer Erbium-Doped Fiber compared to the C-band EDFA.

The C + L-band EDFA was developed in the early 2000s for the significant increase in capacity in WDM systems, initially for terrestrial transmission systems. However, L-band EDFAs have not been used in submarine cable systems due to their lower energy efficiency compared to C-band EDFAs, and the limited power availability in submarine systems. Despite this, the necessity for L-band amplification in trans-oceanic systems has been increasing. For example, the Pacific Light Cable Network (PLCN) introduced C + L-band EDFA [5]. This system is capable of achieving more than double the fiber capacity of the initial design of FASTER, with a potential cable capacity of up to 144 Tb/s.

To further expand the amplification bandwidth of optical amplifiers, the use of Raman amplifiers based on the Raman effect in optical fibers has also been explored. In Raman amplifiers, when a pump light is injected into the fiber, gain occurs in the wavelength range that is approximately 100 nm longer than the pump light's wavelength. By selecting the wavelength of the pump light, amplifiers for arbitrary wavelength bands can be achieved. Moreover, by using multiple pump light with different wavelengths, wideband optical amplifiers can also be created [6]. As shown in Fig. 8.14, hybrid amplification using both EDFA and Raman amplifier is also possible. For example, the Argentina-Brazil submarine cable (ARBR) uses a hybrid amplification system combining C-band EDFA and L-band Raman amplification [7]. This system provides a gain bandwidth of around 70 nm, covering both the C-band and L-band. Unlike the C + L-band EDFA, there is no need for a guard band between the bands, resulting in a continuous amplification bandwidth. This characteristic is becoming increasingly important as the demand for higher capacity in trans-oceanic cable systems rises.

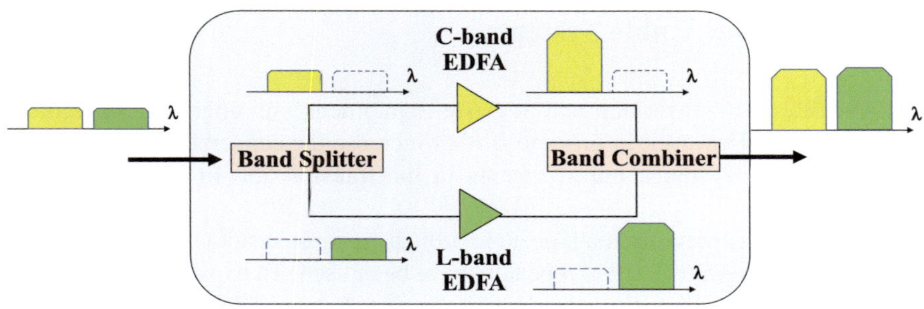

Fig. 8.13 Configuration of C + L band EDFA and spectra

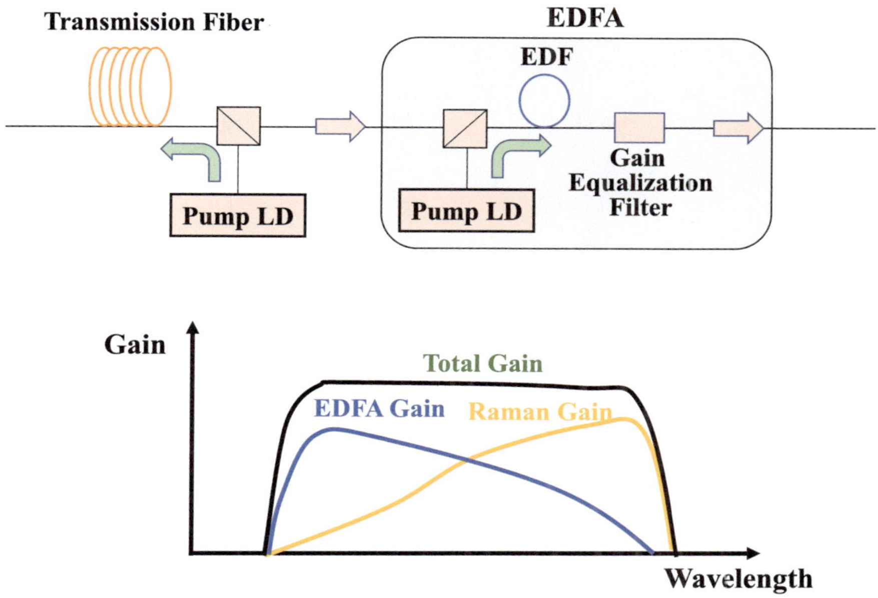

Fig. 8.14 Configuration of hybrid amplification with EDFA and Raman amplifier

8.4 Open Submarine Cable

The digital coherent detection can remove the dispersion management in the transmission line. One of the important practical aspects of such system is weak dependence between the optimization of the transmission line and the transponder. Based on this, the open cables concept was introduced in submarine cable systems, proposing that the transmission line can be optimized independently of the transponder technology [8]. This approach enables system designs that support flexible transponder selection, as long as defined performance requirements are satisfied. Open cables have driven a significant industry transition, offering major capacity gains through the independent selection of the best components both in the transmission line and transponders. They offer flexibility, allowing the cable operators to deploy the best-of-breed coherent transponders over the cable's lifetime. Each generation of transponder brings advantages such as higher spectral efficiency, smaller footprint, and lower power consumption and cost.

The open cables concept has required a redefinition of the performance metrics and acceptance criteria historically applied to submarine cable systems, which were traditionally the capacity and Q-margin of a specific transponder technology. Gaussian or Generalized Signal-to-Noise Ratio (GSNR) were proposed as a performance metric, based on the fact that fiber nonlinear effects can be considered as an additive white Gaussian noise (AWGN) statistically independent of ASE noise in optical fiber transmission without dispersion management [9]. GSNR can be calculated with the

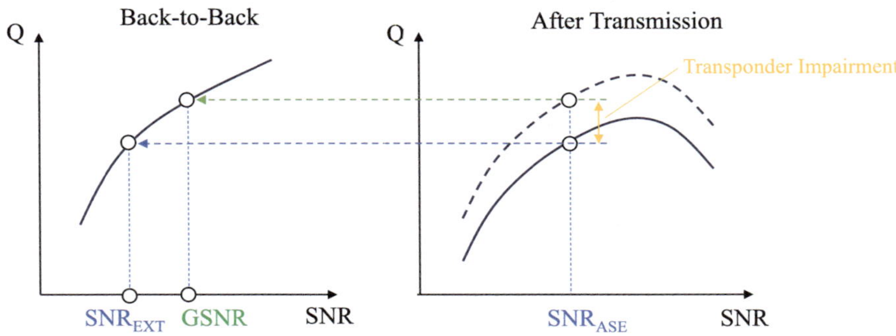

Fig. 8.15 GSNR measurement method

following equation.

$$\frac{1}{\mathrm{GSNR}} = \frac{1}{\mathrm{SNR}_{\mathrm{ASE}}} + \frac{1}{\mathrm{SNR}_{\mathrm{NLI}}} \tag{8.4.1}$$

where $\mathrm{SNR}_{\mathrm{NLI}}$ is calculated with the noise coming from fiber nonlinearities. GSNR is a metric independent on modulation format and channel spacing.

The GSNR of a transmission line is obtained according to the following procedure.

(1) In back-to-back configuration with a pair of test transponders, Q-factor performance (back-to-back curve) as shown in the left plot in Fig. 8.15 is measured with changing SNR by adding ASE noise.
(2) The same transponder pair is connected to the submarine transmission line and Q-factor after transmission is obtained.
(3) Using the obtained Q-factor after transmission and the back-to-back curve, $\mathrm{SNR}_{\mathrm{EXT}}$ is obtained as shown in Fig. 8.15.
(4) GSNR is finally obtained by subtracting the contribution of the transponder impairment:

$$\frac{1}{\mathrm{GSNR}} = \frac{1}{\mathrm{SNR}_{\mathrm{EXT}}} - \frac{1}{\mathrm{SNR}_{i}} \tag{8.4.2}$$

SNR_i includes the noise caused by the transponder and is generally estimated with simulated or laboratory-based measurements. The obtained GSNR can be used as a transponder-independent metric of the performance of the transmission line.

To obtain a more accurate GSNR, other effects such as Guided-Acoustic-Wave Brillouin Scattering (GAWBS) and signal droop must also be considered.

GAWBS is a spontaneous scattering of light in an optical fiber due to the interaction of transverse vibrational acoustic modes with light [10]. The impact of GAWBS became more pronounced in high bit rate digital coherent detection systems due to the introduction of low loss optical fibers. Since GAWBS is a narrowband effect which scatters light within channel bandwidth itself, it can't be measured directly and accounted for in the process of typical OSNR system characterization. Therefore, GAWBS SNR penalty is usually estimated analytically.

In submarine cable systems, the optical amplifiers in repeaters are usually operated in constant output power mode. In this case, the signal power is depleted with large noise accumulation. This effect is referred to as signal droop. The effect of signal droop should be accurately accounted for, especially in high-capacity submarine cable systems with large number of fiber pairs, where the fiber launch power in each fiber becomes lower.

Problems

8.1 Coherent detection in optical communication systems was intensely studied in 1980s, but not commercialized before 2000. Explain its reason.

8.2 Explain the difference of QAM signals with Geometric Shaping and Probabilistic Shaping compared with the conventional QAM.

8.3 Explain the technologies to increase the signal bandwidth further than C-band in WDM transmission systems.

References

1. S. Tsukamoto, K. Katoh, K. Kikuchi, Unrepeated transmission of 20-Gb/s optical quadrature phase-shift-keying signal over 200-km standard single-mode fiber based on digital processing of homodyne-detected signal for group-velocity dispersion compensation. IEEE Photon. Technol. Lett. **18**(9), 1016–1018 (2006)
2. H. Sun, K.-T. Wu, K. Roberts, Real-time measurements of a 40 Gb/s coherent system. Opt. Express **16**(2), 873–879 (2008)
3. V. Kamalov, L. Jovanovski, V. Vusirikala, E. Mateo, Y. Inada, T. Ogata, K. Yoneyama, P. Pecci, D. Seguela, O. Rocher, H. Takahashi, FASTER open submarine cable, in *ECOC2017*, Th2E.5 (2017)
4. V. Kamalov, L. Jovanovski, V. Vusirikala, S. Zhang, F. Yaman, K. Nakamura, T. Inoue, E. Mateo, Y. Inada, Evolution from 8QAM live traffic to PS 64-QAM with neural-network based nonlinearity compensation on 11000 km open subsea cable, in *OFC2018*, Th4D.5 (2018)
5. http://pldcglobal.com/plcn-is-coming-in-service-in-2019/
6. S. Namiki, Y. Emori, Ultrabroad-band Raman amplifiers pumped and gain-equalized by wavelength-division-multiplexed high-power laser diodes. IEEE J. Sel. Top. Quantum Electron. **7**, 3–16 (2001)
7. https://www.xtera.com/2018/05/01/seaborn-selects-cl-band-design-arbr-serve-needs-icpcom munity/
8. E. Hartling, A. Pilipetskii, D. Evans, E. Mateo, M. Salsi, P. Pecci, P. Mehta, Design, acceptance and capacity of subsea open cables. IEEE J. Lightwave Technol. **39**(3), 742–756 (2021)
9. P. Poggiolini, G. Bosco, Carena, V. Curri, Y. Jiang, F. Forghieri, The GN-model of fiber non-linear propagation and its applications. IEEE J. Lightwave Technol. **32**(4), 694–721 (2014)
10. R. Shelby, M. Levenson, P. Bayer, Guided acoustic-wave Brillouin scattering. Phys. Rev. B **31**(8), 5244–5252 (1985)

Chapter 9
Ultra-High-Capacity Transmission Using Space Division Multiplexing

Abstract As described in Chap. 7, the transmission capacity of optical fiber transmission systems has dramatically increased with the introduction of WDM technology. However, even with the adoption of digital coherent reception technology explained in Chap. 8, it has become evident that this capacity increase is approaching its limit. To overcome this limitation, the introduction of new innovative technologies is essential. Space division multiplexing (SDM) technology using multi-core fibers (MCF) or multi-mode fibers (MMF) is a promising candidate to overcome this limitation. This chapter provides an overview of SDM technology and introduce large-capacity transmission experiments utilizing this approach.

Keywords Space division multiplexing · Multi-core fiber · Multi-mode fiber · Few-mode fiber · Core-to-core crosstalk · Multiple Input Multiple Output

9.1 Capacity Limit in Optical Fiber Communication Systems

Communication traffic has been continuously increasing at an annual rate of approximately 40%. Assuming traffic continues to grow at the same pace, it is expected to reach 100 times the current level by the latter half of the 2020s. Consequently, it will be necessary to expand the fiber capacity to the 1 Pb/s level in backbone optical transmission systems.

However, achieving such high capacity is challenging with conventional optical fibers for the following reasons.

The maximum theoretical value of communication capacity (Shannon limit), derived from the Shannon–Hartley theorem, is expressed by the following equation:

$$C\ [\mathrm{b/s}] = B\ [\mathrm{Hz}]\log_2(1 + \mathrm{SNR}) \tag{9.1.1}$$

Here, C represents the communication capacity, B denotes the bandwidth, and SNR is the signal-to-noise ratio. Rewriting the formula in terms of spectral efficiency

M. Suzuki and I. Morita, *Long-Haul Optical Fiber Communication Systems*, Textbooks in Telecommunication Engineering, https://doi.org/10.1007/978-3-032-19472-5_9

(SE), which represents the communication capacity per frequency, it becomes the following equation:

$$\text{SE [b/s/Hz]} = \log_2(1 + \text{SNR}) \quad (9.1.2)$$

Before the introduction of digital coherent detection, the modulation formats which could be used in optical fiber communication systems were limited to those that utilized either the amplitude or the phase of light. In this case, even when employing a four-level modulation format such as DQPSK (see Chap. 3), it was impossible to achieve a spectral efficiency exceeding 2 b/s/Hz, regardless of how high the SNR was made. With the introduction of digital coherent detection, it became possible to employ modulation formats that utilize both the amplitude and phase of light, enabling spectral efficiencies close to the theoretical limit given by Eq. 9.1.2.

Equations 9.1.1 and 9.1.2 indicate that if the SNR can be increased with higher signal power, the capacity, and the spectral efficiency can be increased accordingly. In optical fiber communication systems, however, it is necessary to consider fiber nonlinearity in light propagation through the fiber. In optical fiber communication systems, optical signal power is concentrated within a core with a diameter of approximately 10 μm, resulting in a very high optical power density. This makes the effects of fiber nonlinearity significant, especially over long transmission distances.

Figure 9.1 schematically illustrates the Shannon limit derived from Eq. 9.1.2 and the effects of fiber nonlinearity. As shown in the figure, due to the impact of fiber nonlinearity, there is an upper limit to the achievable spectral efficiency. Furthermore, as the transmission distance increases, the influence of fiber nonlinearity grows, causing this upper limit to decrease. Considering parameters for standard single-mode fiber (SMF) with an effective core area of 80 μm^2 and a loss of 0.2 dB/km, and assuming ideal distributed Raman amplification, it has been reported that the spectral efficiency per polarization is limited to approximately 9.0 and 5.6 b/s/Hz for 500 and 8,000 km transmission, respectively [1]. In this report, the spectral efficiency is calculated for a single polarization, and if polarization multiplexing is applied, the spectral efficiency can be doubled. If the entire bandwidth of the C+ L-band commonly used in optical fiber communication (about 10 THz) can be utilized, the transmission capacity limit is estimated to be approximately 100–200 Tb/s.

In addition to fiber nonlinearity, there is another phenomenon called "fiber fuse," which limits the fiber launch power. This phenomenon occurs when high-power light in the watt range is launched into an optical fiber, and a portion of the fiber is heated, causing only the core to partially melt. The melted region emits intense light as it propagates back toward the light source, leaving a longitudinal cavity in the core area of the fiber after the molten region passes through. In other words, the optical fiber's structure is physically destroyed.

It should be noted that fiber fuse does not always occur when high power is input. Instead, it begins when a cause for core melting arises at a specific point, propagating from there toward the light source. In actual optical fiber transmission systems, potential causes include the absorption of light by contaminants adhered to

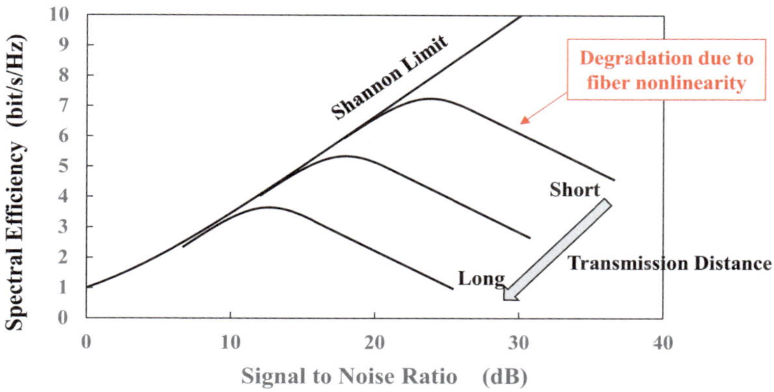

Fig. 9.1 Spectral efficiency obtained in optical transmission systems

connector end-faces or the concentration of optical energy due to multiple reflections when a fiber breaks.

There is a threshold power level at which fiber fuse occurs. This threshold is proportional to the mode field diameter of the optical fiber. For standard SMF, the threshold is reported to be approximately 1.5 W. In a 240 km transmission experiment using standard SMF that achieved a transmission capacity of 69 Tb/s, the total fiber launch power reached 120 mW [2]. To further increase transmission capacity and extend transmission distance, it is necessary to further increase the fiber launch power. However, as the input power approaches the fiber fuse threshold, this limitation introduces a constraint on the transmission capacity achievable with a single optical fiber.

Thus, considering the fiber nonlinearity and the fiber fuse, the primary limitation of fiber launch power makes it impossible to achieve a transmission capacity of 1 Pb/s or more, which is 100 times the current level, using conventional SMF. Therefore, as a means to overcome the limitations of conventional optical fibers, there is growing interest in exploring space division multiplexing (SDM) optical transmission systems.

9.2 Outline of Space Division Multiplexing Transmission Systems

In the SDM optical transmission systems, significant increase of the fiber capacity can be expected by using multi-core fibers (MCFs) or multi-mode fibers (MMFs). This is achieved by transmitting independent data through each "core" or "mode."

Figure 9.2 illustrates the optical fibers used in SDM transmission systems. In conventional optical fibers, a single core is provided within the fiber, and only a single-mode is transmitted (single-core, single-mode). In contrast, MCFs have multiple cores embedded within the optical fiber. Depending on the crosstalk between cores, MCFs can be classified as coupled or uncoupled MCF. Uncoupled MCFs have negligible crosstalk between cores and can be treated similarly to multiple independent optical fibers. Coupled MCFs, on the other hand, require digital signal processing at the receiver to demultiplex signals for treating each core as an independent transmission channel.

MMF, meanwhile, is an optical fiber designed with a core diameter larger than that of a single-mode fiber, allowing multiple modes to propagate. MMF was developed before single-mode fibers and was used in early optical transmission systems, where a single data was transmitted with all modes without distinguishing between them. In SDM transmission systems, MMF expands transmission capacity by treating each propagating mode individually and transmitting independent data through each mode. When the number of propagating modes within the fiber is less than around ten, it is also referred to as a few-mode fiber (FMF). FMF is designed to allow the desired modes to propagate by appropriately setting the core diameter and refractive index profile.

In the case of MMF, coupling between modes can be suppressed by designing the core shape as elliptical. However, the transmission distance for which mode coupling can be suppressed is limited, typically to about 0.5 km or less. In standard MMF, mode coupling within the fiber cannot be entirely avoided. Therefore, to transmit independent data through each mode, digital signal processing at the receiver is essential for mode separation.

Furthermore, multi-mode, multi-core fibers (MM-MCF), which make each core of an MCF multi-mode, are also used as optical fibers for SDM transmission. In this case, the transmission capacity can be increased by a factor of the product of the number of cores and modes, making MM-MCF an attractive option for ultra-high-capacity optical transmission.

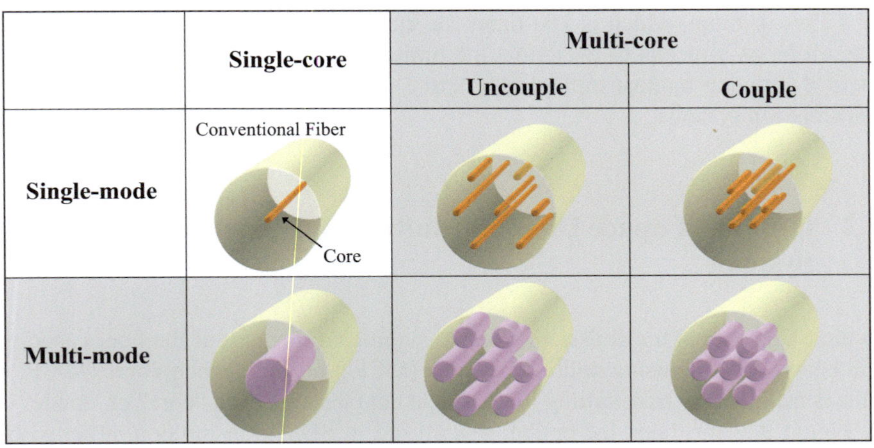

Fig. 9.2 Optical fibers used in SDM transmission systems

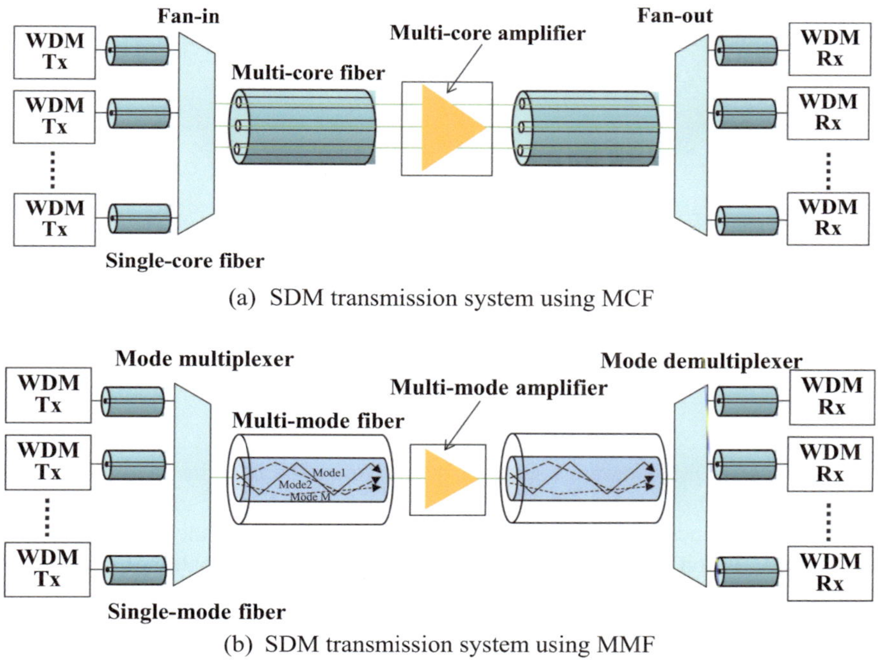

Fig. 9.3 Basic configuration of SDM transmission systems

Figure 9.3 shows the configuration of an SDM transmission system. In SDM transmission systems using MCF, the transmitter and receiver employ fan-in and fan-out (FIFO) to connect the multiple cores of an MCF to multiple single-core optical fibers. The FIFO must exhibit uniform characteristics across the entire wavelength band being utilized and various designs have been explored, including fiber bundle types [3], space coupling types [4], and 3D waveguide types [5].

On the other hand, in SDM transmission systems using MMF, the transmitter uses a mode multiplexer. This device converts the signal light in the LP_{01} mode, output from transmitters prepared for each mode, into different modes (e.g., LP_{11a}, LP_{11b}), and then multiplexes them. At the receiver, a mode demultiplexer with inverse characteristics to the mode multiplexer separates the mode-multiplexed signal light into its constituent modes and then converts each into the LP_{01} mode for output. Several types of mode multiplexers and demultiplexers have been studied, including photonic lantern types [6], spatial optical types using phase plates [7], and fiber coupler types [8].

For long-distance transmissions, such as backbone transmission systems spanning hundreds of kilometers or more, repeater transmission using optical amplifiers is indispensable. In MCF transmission systems, it is possible to configure the system by using multiple single-core optical amplifiers individually for each core. However, this approach fails to achieve space-saving benefits, which is one of the significant

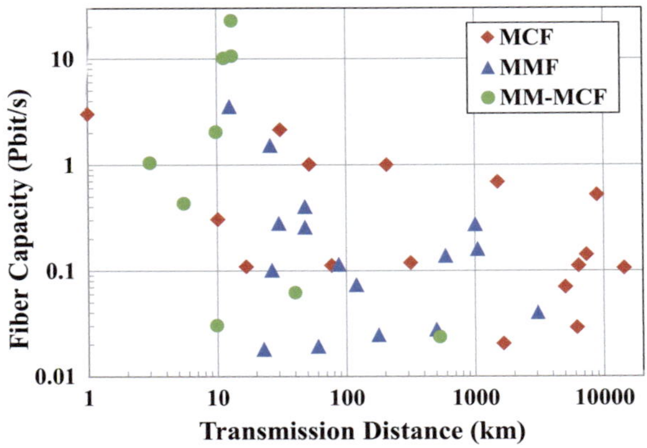

Fig. 9.4 Large-capacity transmission experiments using SDM

advantages of SDM systems. Therefore, to maximize the benefits of long-distance transmission systems using MCF, it is essential to implement multi-core optical amplifiers to reduce both space and power consumption. On the other hand, for MMF transmission systems, a multi-mode optical amplifier capable of amplifying all modes used in transmission equally is required.

Figure 9.4 shows major experimental results of high-capacity SDM transmission using MCF, MMF, and MM-MCF, each achieving transmission capacities of 10 Tb/s or more. In SDM transmission using MCF, long-distance transmission exceeding 10,000 km has been achieved with high-capacity transmissions exceeding 100 Tb/s. In SDM transmission using MMF, high-capacity transmissions exceeding 100 Tb/s have also been achieved; however, the transmission distance is limited to around 1000 km, highlighting the challenge of further extending the transmission distance. In SDM transmission using MM-MCF, transmission experiments achieving capacities of higher than 10 Pb/s have been reported [9–11].

Moreover, the SDM transmission offers advantages not only in terms of capacity expansion but also in power efficiency.

When SDM transmission with M channels is introduced, it is possible to increase the communication capacity M-fold within the same bandwidth. Consequently, the maximum theoretical value of spectral efficiency derived from the Shannon-Hartley theorem is expressed by the following equation:

$$SE_{SDM}\ [\text{b/s/Hz}] = M \log_2(1 + \text{SNR}) \tag{9.2.1}$$

Using Eqs. 9.1.2 and 9.2.1, the calculated spectral efficiency in cases without SDM and with 2-channel and 4-channel SDM are shown in Fig. 9.5.

In the case without SDM, to double the spectral efficiency from 1 to 2 b/s/Hz, a 4.8 dB increase in the SNR is required. To further double the spectral efficiency from 2 to 4 b/s/Hz, a 7 dB increase in the SNR is necessary. This indicates that doubling the spectral efficiency requires more than double the optical signal power. Specifically, to increase the spectral efficiency from 1 to 2 b/s/Hz, three times the optical signal

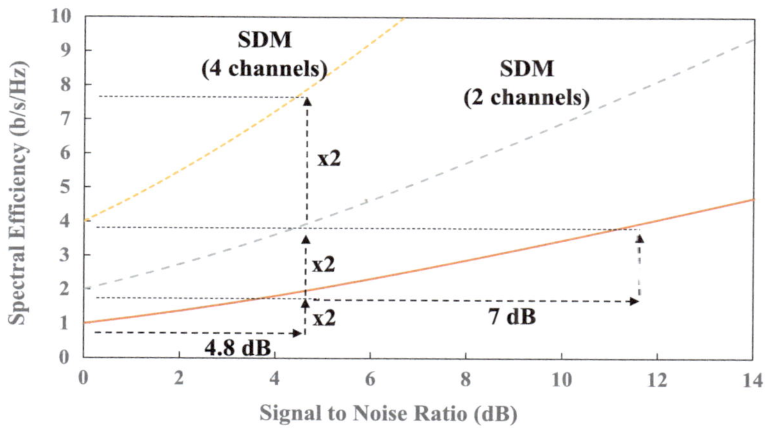

Fig. 9.5 Relation between SNR and spectral efficiency

power is required, and to increase it from 2 to 4 b/s/Hz, five times the optical signal power is needed.

On the other hand, with SDM the spectral efficiency can be increased by the number of spatial channels, and the required SNR per spatial channel remains constant. This indicates that the optical signal power only needs to be increased in proportion to the number of spatial channels, meaning that to double the spectral efficiency, the required optical signal power is also doubled.

Especially when increasing the capacity of optical transmission systems, the increase in required energy is an unavoidable issue, and from this perspective, SDM transmission is highly anticipated.

In the following sections, SDM transmission using MCF, MMF, and MM-MCF is explained.

9.3 Space Division Multiplexing Transmission Using Multi-core Fibers

9.3.1 Long-Haul Transmission Using Uncoupled MCF

In SDM transmission using uncoupled MCF, it is necessary to sufficiently suppress core-to-core crosstalk so that each core can be regarded as an independent transmission path. To increase capacity, it is desirable to increase the number of cores in the MCF. However, to suppress core-to-core crosstalk while maintaining a sufficient spacing between the cores, the cladding diameter must be increased. Since excessively increasing the cladding diameter is not feasible from a mechanical reliability standpoint, it is important to suppress core-to-core crosstalk even when the core spacing is reduced.

One solution to this problem is the introduction of a trench structure into the MCF. As shown in Fig. 9.6, this involves creating a trench around the core area with a lower refractive index. By adding such a trench, the signal distribution can be concentrated around the core, reducing the overlap of the signal tails from each core and thereby suppressing core-to-core crosstalk.

Another solution to suppress core-to-core crosstalk is to make the propagation constants of each core different. In this method, the refractive index distribution of the cores that are most adjacent to each other, where core-to-core crosstalk is maximized, is made uneven so that the propagation constants differ. A transmission experiment using a 12-core MCF with such a core arrangement, as shown in Fig. 9.7, has been reported where crosstalk between cores can be effectively suppressed by using two types of cores [12].

In long-distance optical transmission systems, optical amplifiers are indispensable to compensate for the loss of transmission fiber. In MCF transmission systems, from the perspective of miniaturization and energy efficiency, it is desirable to use optical

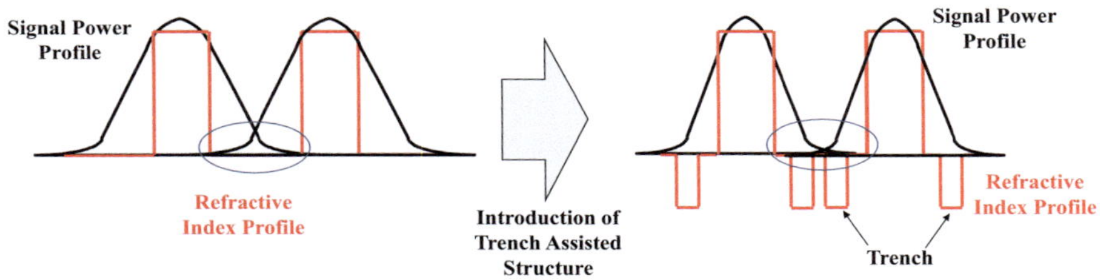

Fig. 9.6 Introduction of trench-assisted structure

Fig. 9.7 Inhomogeneous MCF

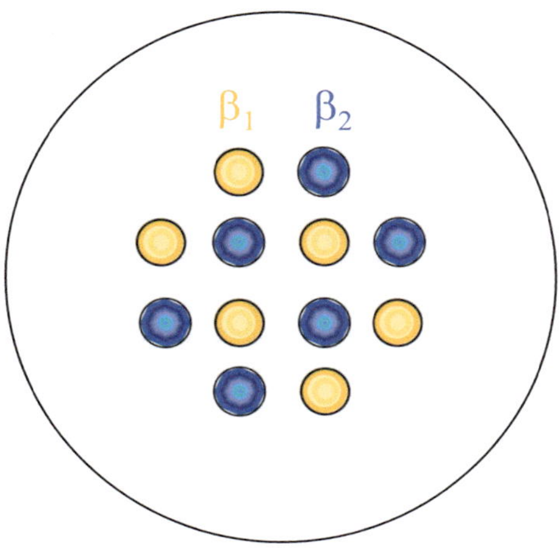

amplifiers with muti-core Erbium-Doped Fiber (EDF). Multi-core EDF (MC-EDF) amplifies are classified into two types based on how the pump light is input into the MC-EDF.

One type is the individual-core-pumped multi-core optical amplifier (Fig. 9.8a), where pump light is input individually to each core of the MC-EDF through fan-in/ fan-out components. The other type is the clad-pumped multi-core optical amplifier (Fig. 9.8b), where pump light is input into the cladding part of the MC-EDF with a double-clad structure, pumping multiple cores simultaneously. The clad-pumped multi-core optical amplifier has the advantage of reducing the number of components and simplifying the structure because all cores share the pump light. However, the challenge lies in controlling the gain characteristics of each core and pumping efficiency. On the other hand, the individual-core-pump multi-core optical amplifier requires separate components, such as pump light sources, for each core, which makes the structure more complex as the number of cores increases. However, it has the advantage of being able to control the gain for each core independently.

In a transmission experiment using an individual-core-pumped multi-core optical amplifier, a transmission capacity-distance product of 1.03 Exa b/s km was achieved [13]. Figure 9.9 shows the experimental setup. The system involved a 319 km

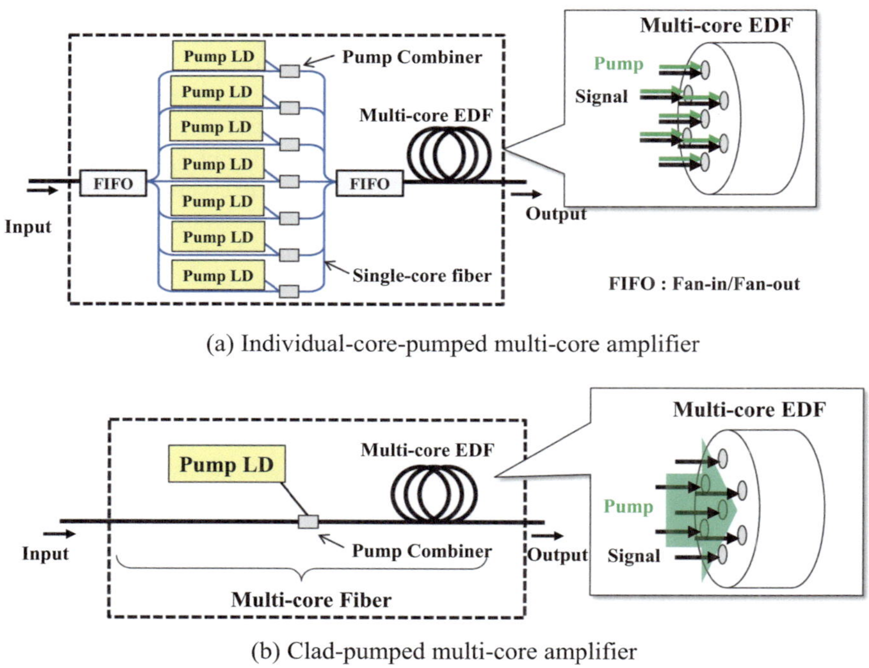

Fig. 9.8 Multi-core amplifiers

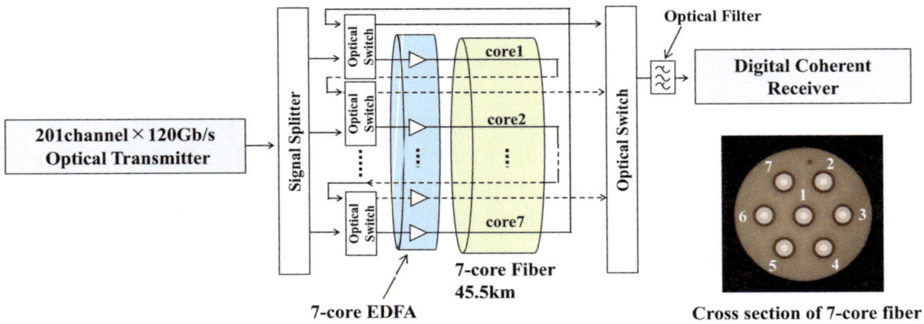

Fig. 9.9 Experimental setup using individual-core-pumped 7-core amplifiers

transmission link with 7-core MCF, where each core was connected through a 7-core optical amplifier. The long-distance transmission characteristics were evaluated through this recirculating loop experiment with a span length of 45.5 km.

For the transmitted signal, a 120 Gb/s (symbol rate of 30 Gbaud) polarization-multiplexed QPSK signal was used, with the optical signal undergoing Duo-Binary encoding to compress the bandwidth. This bandwidth compression allowed the modulated signals to be wavelength-multiplexed at a frequency spacing smaller than the symbol rate, 25 GHz. The sensitivity penalty due to Duo-Binary encoding was compensated using maximum likelihood sequence estimation at the receiver. The experimental setup assumed a soft-decision forward error correction (FEC) code with a redundancy of 20%, giving an effective bit rate of 100 Gb/s after excluding the overhead. The 201 WDM signals were input into each core of the MCF via a signal splitter and optical switches. The optical switches were controlled to simulate the condition where each core receives an individual optical signal.

The 7-core fiber used in the transmission experiment introduced a trench structure to suppress core-to-core crosstalk. After 45.5 km of optical fiber, the core-to-core crosstalk was less than − 51 dB for all cores. The multi-core optical amplifier used in this experiment was a two-stage 7-core EDFA, with a maximum output power of 23 dBm per core and a noise figure below 5.5 dB for all cores.

At the optical receiver, a wavelength filter was used to select the measured channel from the WDM signals. The signal was then received by a digital coherent receiver. After analog-to-digital conversion, the data was stored, and digital signal processing was performed offline.

The transmission performance of all channels was evaluated, and it was confirmed that transmission performance over the FEC threshold was achieved after 7,326 km. The transmission capacity-distance product obtained in this experiment was 1.03 Exa b/s km (transmission capacity 140.7 Tb/s × transmission distance 7,326 km), making this the first transmission experiment to exceed a capacity-distance product of 1 Exa b/s km.

Transmission experiments using multiple single-core optical amplifiers for each core in a MCF were also reported [14, 15]. In these experiments, a 12-core MCF was used as shown in Fig. 9.10. Each core's signal was input into individual single-core optical amplifiers via FIFO and amplified separately. To achieve power savings, a single pump light source was shared by the 12 optical amplifiers. Furthermore, the bandwidth for WDM transmission using C-band was limited to about 22 nm, which minimized the energy loss caused by gain equalization filters, contributing to further power savings. As a result, it became possible to pump 12 single-core optical amplifiers using only a 800 mW pump light source. Using this type of repeaters, 105.1 Tb/s signals were successfully transmitted over 14,350 km [14]. In another experiment using the C + L band, the bandwidth was extended to 73 nm, and multi-level modulation was employed, increasing the transmission capacity to 520 Tb/s. The transmission over 8,830 km was also successfully demonstrated [15]. The transmission capacity-distance product obtained in this experiment reached 4.59 Exa b/s km, marking the highest value achieved to date.

Transmission experiments using a clad-pumped multi-core optical amplifier have been reported, demonstrating a total transmission capacity of 70 Tb/s (7 cores × 50 wavelengths × 200 Gb/s) using a 7-core MCF [16].

Figure 9.11 illustrates the configuration of the clad-pumped multi-core optical amplifier used in the transmission experiment. To efficiently pump the amplifier, the 7-core EDF used a double-clad structure. The first clad made of silica glass was surrounded by a second clad made of polymer material. The pump light, with a wavelength of 976 nm, was input into the first clad from a multi-mode laser. By inputting 42.3 dBm pump light into the 50-m EDF, an average gain of larger than

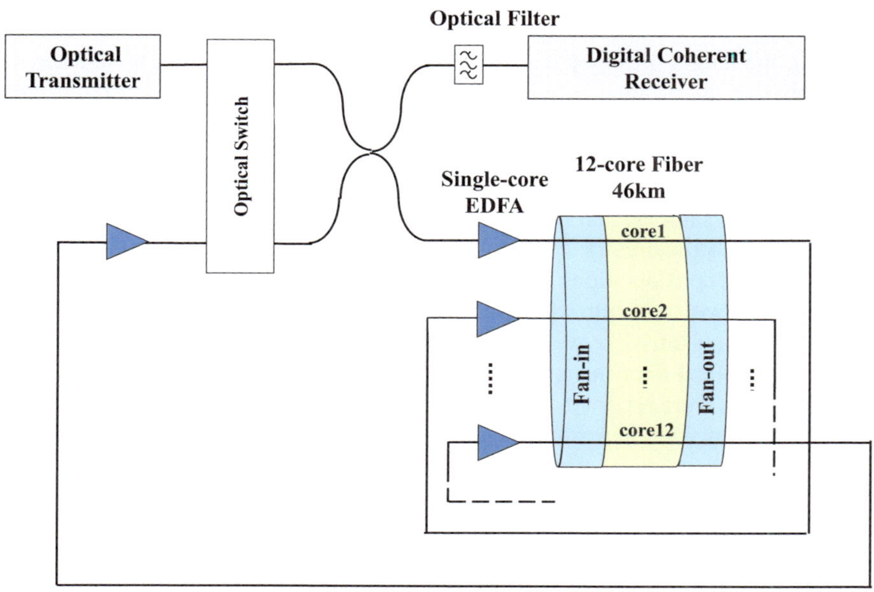

Fig. 9.10 Experimental setup for MCF transmission using single-core optical amplifiers

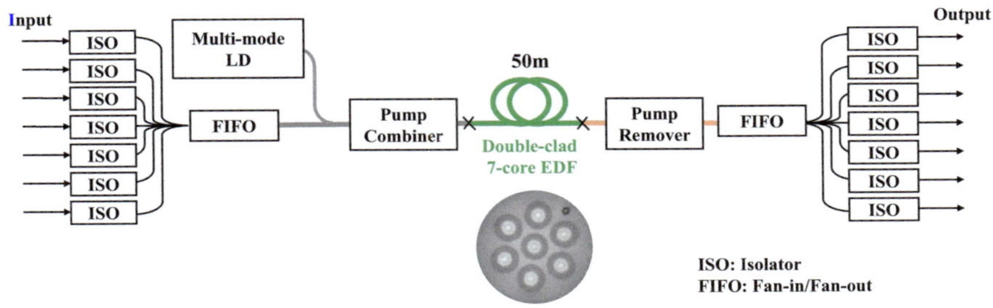

Fig. 9.11 Clad-pumped multi-core amplifier

13 dB and a noise figure of lower than 5.8 dB were achieved in the wavelength range of 1,570–1,610 nm at an input signal power of 7.5 dBm [17].

In the transmission experiment, 50 channels of 64 Gbaud polarization-multiplexed QPSK signals (bit rate: 256 Gb/s) were wavelength-multiplexed at 70 GHz spacing over a wavelength range of 1578.3–1600.5 nm. The long-distance transmission performance was evaluated in recirculating loop experiments using a 40 km span of 7-core fiber. As a result, after 5,040 km of transmission, it was confirmed that the transmission performance exceeded the threshold of the soft-decision FEC code with a redundancy of 25.5% across all channels.

In the aforementioned transmission experiments, MCFs with larger cladding diameters than the standard SMF (125 μm) were used to increase the number of cores. However, for practical reasons, it is preferable to maintain a cladding diameter of 125 μm. If the cladding diameter were to be doubled (e.g., from 125 to 250 μm), the length of the optical fiber that can be produced would be reduced to one-quarter of the original length. Furthermore, in terms of cable formation (e.g., wire coloring, tape core wiring, or integration into cable structures), if the optical fiber diameter remains the same, existing equipment can be reused. Additionally, in terms of connections, while alignment along the axis is crucial for MCF connections, maintaining the same optical fiber diameter allows for the continued use of existing connectors and structures.

Given these considerations for the early practical implementation of MCF, 4–5 core MCFs with 125 μm cladding diameters were studied. With such MCFs, high-capacity transmission over 100 Tb/s with interconnected MCFs manufactured by different optical fiber manufacturers [18] and ultra long-haul transmission over 12,000 km [19] were reported.

In the transmission experiment using interconnected MCFs [18], 4-core MCFs produced by three optical fiber manufacturers with the same specifications were intentionally interconnected and configured three transmission spans ranging from 104–107 km as shown in Fig. 9.12. The fabricated 4-core MCFs satisfied the standard SMF specifications such as cutoff wavelength, mode field diameter, bending loss, and polarization mode dispersion. The loss in each transmission span was compensated for by a cladding-pumped multi-core optical amplifier. In this configuration, the total transmission length was 316 km. 36 Gbaud polarization-multiplexed 16QAM signals

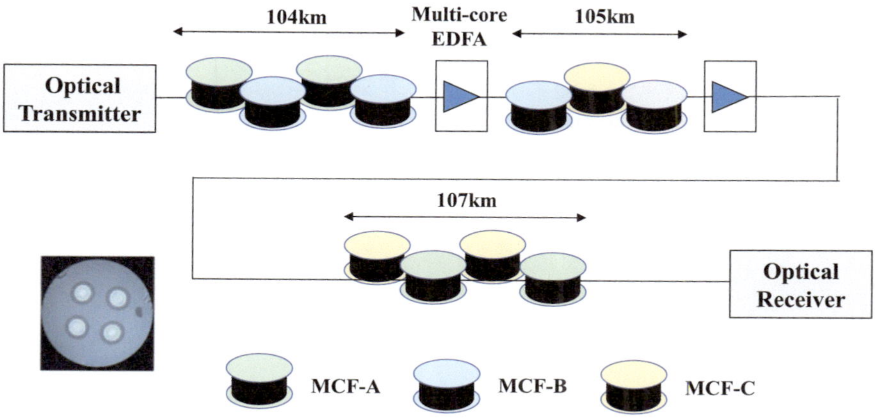

Fig. 9.12 Interconnection transmission experiment using 4-core MCFs

were wavelength-multiplexed at 37.5 GHz spacing and the total number of channels was 464 with 116 wavelengths and 4 cores. The transmission performance for all channels exceeded the threshold of the soft-decision FEC code with a redundancy of 12.75%. This result demonstrated that a transmission capacity of 118.5 Tb/s could be achieved using MCFs with standard cladding diameters.

The ultra-long-haul transmission experiments using uncoupled 4-core fibers with standard cladding diameter were conducted using the experimental setup shown in Fig. 9.13 [19]. Transmission performance was evaluated with 24-Gbaud polarization multiplexed QPSK signals with 25 GHz frequency spacing. The major characteristics of the 4-core fiber at 1550 nm are summarized in Table 9.1. The span loss including the loss of FIFO was compensated by single-core EDFAs. Q-factors of all SDM and all WDM channels after 9,150 and 12,040 km were measured, and it was confirmed that fiber capacity of 63.17 Tb/s over 9150 km and 55.94 Tb/s over 12,040 km could be achieved with uncoupled 4-core fibers with standard cladding diameters.

In 2023, it was announced that the adoption of uncoupled 2-core fiber on some portion in Taiwan-Philippines-U.S. cable system, which will be ready for service in 2026 [20]. Although the number of cores is limited to two, this is the first commercial use of MCF in a submarine cable system. Following it, the further introduction of MCF in submarine cable systems is expected.

9.3.2 Long-Haul Transmission Using Coupled MCF

In coupled MCF, there is no need to increase the core spacing in order to suppress crosstalk between cores as the case of uncoupled MCF. This allows for an increase in the number of cores without enlarging the cladding diameter, offering an advantage

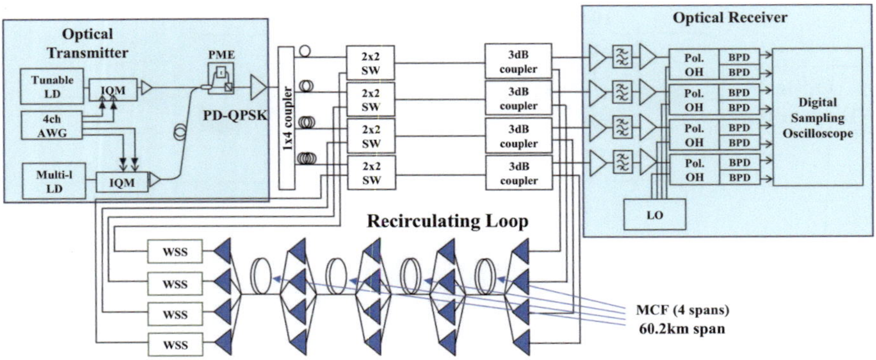

Fig. 9.13 Ultra long-haul transmission experiment using 4-core MCF

Table 9.1 Major optical characteristics of 4-core MCF at 1550 nm

Cross-sectional image	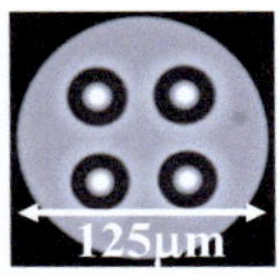
Core-averaged transmission loss	0.156 dB/km
Effective area	87 μm^2
Core pitch	43.0 μm
Cutoff wavelength (7.7 km)	1450 nm
IL of fan-out (FO) devices	0.3–0.5 dB (fibre-bundled type)
Losses at one splice point	< 0.4 dB
Averaged total span losses	10.1 dB
Core-to-core crosstalk with FOs	< – 57.3 dB/span

in spatial utilization efficiency, which is one of the benefits of SDM. However, to mitigate the effects of inter-core coupling that occur during fiber propagation, Multiple Input Multiple Output (MIMO) signal processing, similar to that used in wireless systems, is necessary as shown in Fig. 9.14. This is similar to the case of mode-division multiplexing, which will be discussed in the next section. However, in the case of coupled MCF, the group delay difference between modes is smaller compared to mode-division multiplexing. As a result, the computational load for MIMO signal processing can be reduced. Since the group delay difference between modes accumulates with transmission distance, this characteristic is especially important for long-distance transmission systems, such as submarine cable systems.

Additionally, because there is no need to consider inter-core crosstalk suppression, a simple step-index pure silica core fiber can be used, eliminating the need for

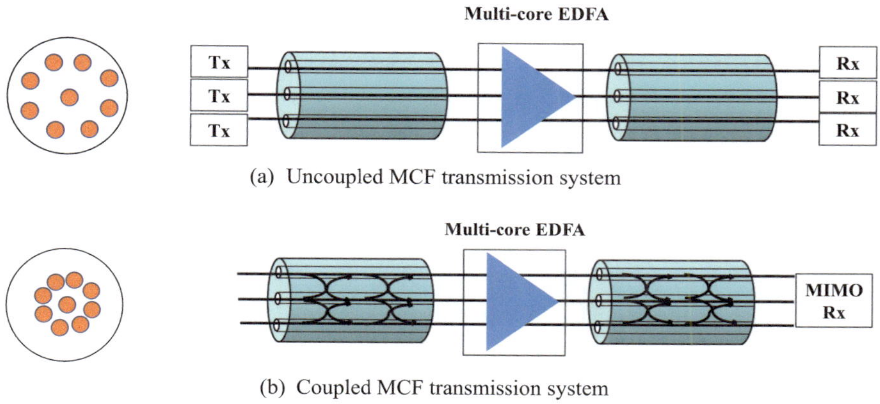

Fig. 9.14 MCF transmission systems

complex refractive index control with Ge doping. As a result, low-loss characteristics similar to those of the ultra-low-loss single-mode fibers recently introduced in submarine cable systems can be expected.

Experiments comparing the performance of coupled MCF and conventional single-core fiber in transoceanic long-distance transmission have been reported [21]. In the experiment, 4-core MCF and 7-core MCF with the same 125 μm cladding diameter as standard SMF were fabricated and compared with SMF with an enlarged effective area to 114 μm^2 using a 120 Gb/s polarization-multiplexed QPSK signal.

The results showed that as the number of cores increased (from 4-core to 7-core), the tolerance to fiber nonlinearity improved. In the case of the 7-core coupled MCF, a 1 dB improvement in performance was observed compared to the effective area enlarged SMF. This indicates that the MCF provides not only an increase in transmission capacity proportional to the number of cores, but also additional benefits.

In this experiment, 8×8 MIMO was used for the 4-core MCF, and 14×14 MIMO was used for the 7-core MCF. The signal processing was carried out offline. For practical implementation, the development of a real-time digital signal processor is indispensable, and reducing the necessary circuit size for such a system is crucial to make this feasible.

9.4 Space Division Multiplexing Transmission Using Multi-mode Fibers

Multi-mode optical amplifiers, just as optical amplifiers for wavelength-division multiplexing (WDM) transmission systems require gain flatness with respect to wavelength, need to amplify all modes equally. An example of propagation modes in MMF is shown in Fig. 9.15.

In conventional single-mode transmission systems, only the LP_{01} mode, which concentrates energy in the center of the optical fiber, propagates at the operating wavelength, so for the optical amplifier, only the LP_{01} mode needed to be considered. However, in mode-division multiplexing, as shown in Fig. 9.15, the electric field distributions differ for each mode, so it is necessary to use a pumping method that can evenly amplify signals with different electric field distributions.

Mode-division multiplexing (MDM) using MMF is broadly classified into two types: uncoupled MDM and coupled MDM. In uncoupled MDM, which suppresses inter-mode coupling during optical fiber propagation by using elliptical cores, the transmission distance is limited to a few kilometers. Therefore, in backbone long-distance transmission systems including optical submarine cable systems, coupled MDM is used.

Coupled MDM is further classified into two types: strongly coupled MDM and weakly coupled MDM, depending on the strength of the inter-mode coupling, as shown in Fig. 9.16.

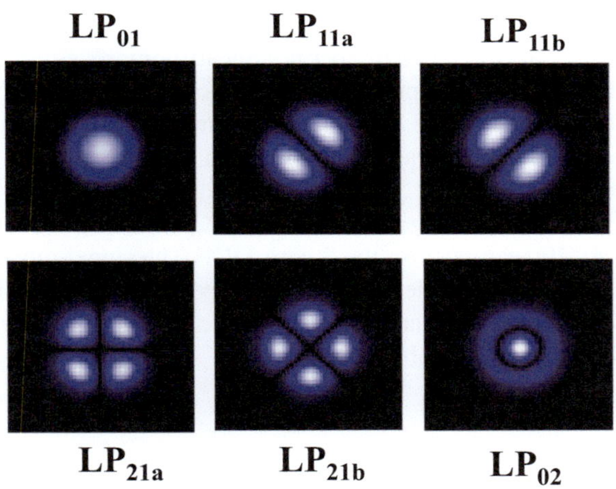

Fig. 9.15 Example of electric field distribution of spatial modes

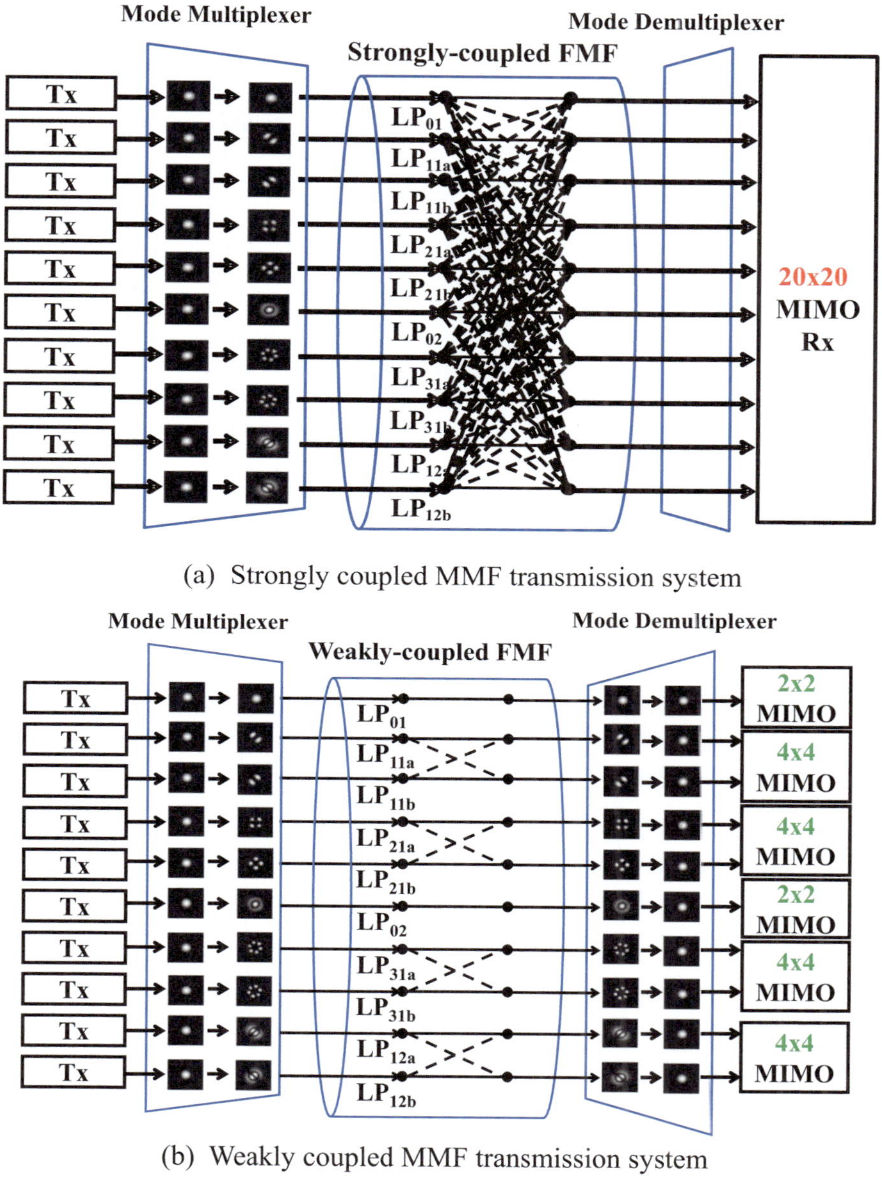

(a) Strongly coupled MMF transmission system

(b) Weakly coupled MMF transmission system

Fig. 9.16 Coupled MMF transmission systems

9.4.1 Strongly Coupled Multi-mode Fiber Transmission

In the strongly coupled MDM, the transmission system is designed under the assumption that mode coupling occurs in the optical fiber, optical amplifier, and mode multiplexers/demultiplexers. The effects of mode coupling that arise in the transmission system are generally suppressed using MIMO signal processing, similar to the coupled MCF transmission described in the previous section. In this system, the strength of mode coupling is not a significant issue, but since individual coherent receivers for each mode must be synchronized, an increase in the number of modes results in increased processing load.

Especially in long-distance transmission, as the transmission distance increases, the differential mode delay (DMD) also increases. To achieve synchronized detection, it is necessary to suppress the effects of DMD, and reducing DMD during optical fiber transmission becomes an important issue. To reduce DMD, methods optimizing the refractive index profile to reduce the inherent DMD of the MMF, as well as using fibers with positive and negative DMDs in combination, similar to dispersion management in conventional WDM systems, have been proposed, as shown in Fig. 9.17.

In the strongly coupled MDM system, the maximum number of modes reported so far is 55 modes [22]. In this experiment, 24.5 Gbaud polarization-multiplexed 16QAM signals were wavelength-multiplexed at 25 GHz spacing. The 25.9 km transmission link consisted of two spools of graded-index MMF, which had a trench-assisted, graded-index profile that supported 55 spatial modes divided into 10 mode groups. The total transmission capacity for the 184-channel WDM transmission was 1.53 Pb/s, with a spectral efficiency of 332 b/s/Hz. This spectral efficiency is the highest achieved in any optical fiber with standard cladding diameter of 125 μm. Additionally, since polarization multiplexing was also used in this report, a 110×110 MIMO signal processing was necessary, but since coherent receivers capable

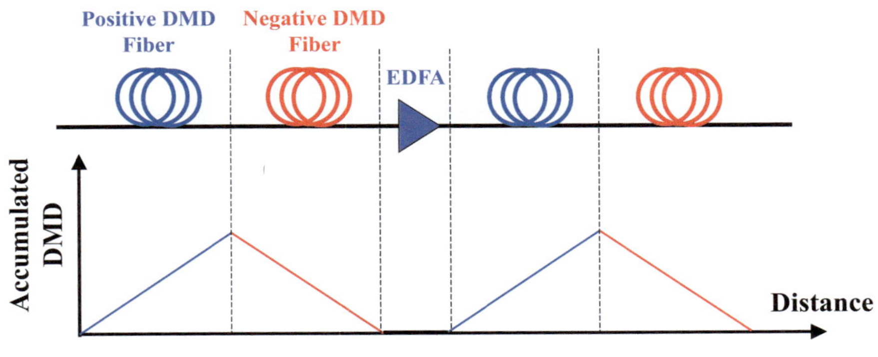

Fig. 9.17 DMD managed transmission line using positive and negative DMD fibers

of receiving all channels simultaneously could not be prepared, time-division multiplexing was employed to simulate synchronized detection with 19 coherent receivers [23].

9.4.2 Weakly Coupled Multi-mode Fiber Transmission

The weakly coupled MDM system is a method where the coupling during the fiber propagation is limited to degenerate modes (e.g., LP_{11a} and LP_{11b}). To suppress the effects of coupling between degenerate modes, MIMO signal processing is required. However, coupling between other modes is sufficiently suppressed, allowing optical channel separation using only a mode multiplexer/demultiplexer. Therefore, even when polarization multiplexing is used, the required MIMO signal processing scale is limited to a maximum of 4 × 4, which is consistent with the MIMO processing used in commercially available 100 Gb/s digital coherent systems, making it more practical from a digital signal processing perspective.

As a transmission experiment using the weakly coupled MDM, 257 Tb/s transmission experiment with a step-index 10-mode fiber has been reported [24]. In this experiment, two carriers of 12 Gbaud polarization-multiplexed QPSK signals were arranged as a channel signal with a bit rate of 96 Gb/s. Such channels were wavelength multiplexed at a 25 GHz spacing in the C and L bands for a 48 km transmission. The transmission line used a 10-mode fiber with a core diameter of 17 μm and a cladding diameter of 125 μm. To suppress mode coupling, the effective refractive index difference between the LP modes was designed to be greater than 6×10^{-4}.

In this transmission experiment, 10 modes were used: LP_{01}, LP_{11a}, LP_{11b}, LP_{21a}, LP_{21b}, LP_{02}, LP_{31a}, LP_{31b}, LP_{12a}, and LP_{12b}, as shown in Fig. 9.18. The mode multiplexing and demultiplexing were performed using a mode multiplexer based on Multiplane Light Conversion (MPLC) technology [7], which generates the desired modes through repeated spatial phase modulation and diffraction.

After transmission, the optical signal was received in two groups using five digital coherent receivers: the first group was the lower-order 5 modes (LP_{01}, LP_{11a}, LP_{11b}, LP_{21a}, LP_{21b}), and the second group was for the higher-order 5 modes (LP_{02}, LP_{31a}, LP_{31b}, LP_{12a}, LP_{12b}). Digital signal processing was performed offline. For the modes without degeneracy (LP_{01}/LP_{02}), 2 × 2 MIMO signal processing was applied, while 4 × 4 MIMO signal processing was applied to the other modes.

In the experiment, an FEC code with a 25.5% redundancy was assumed, and in the case of 336-channel WDM transmission, a total transmission capacity of 257 Tb/s and a spectral efficiency of 31 b/s/Hz were achieved.

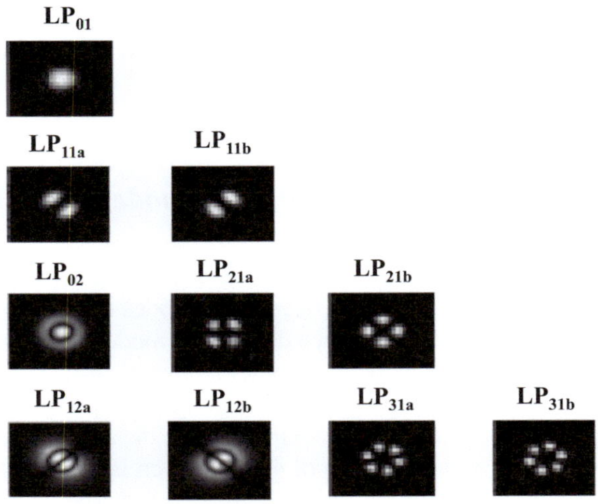

Fig. 9.18 Electric field distribution of modes used in weakly coupled 10 mode transmission experiment

9.5 Space Division Multiplexing Transmission Using Multi-core Multimode Fibers

Multi-core Multi-mode Fiber (MM-MCF) is a type of MCF in which each core is made multi-mode. Its SDM count is the product of the "core count" and the "mode count," making it an effective means for achieving high capacity. Figure 9.19 shows the SDM counts used in SDM transmission experiments. When using MCF and MMF, the SDM counts are typically between 30 and 40. However, by combining both, MM-MCF enables the realization of optical fibers with a SDM count exceeding 100.

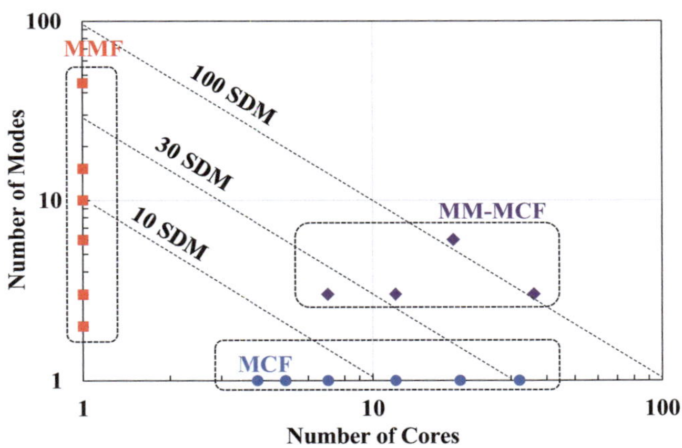

Fig. 9.19 Numbers of cores and modes of fibers used in SDM transmission experiments

Figure 9.20 shows the spectral efficiency and transmission capacity obtained from representative high-capacity transmission experiments. In conventional SMF transmission experiments, spectral efficiencies exceeding 10 b/s/Hz have been achieved, with the maximum value of 17.3 b/s/Hz reported in a transmission experiment using polarization-multiplexed 4,096QAM signals with probabilistic constellation shaping (PCS) [25]. MCF and MMF can achieve SDM counts between 30 and 40, but due to the penalties associated with SDM, transmission experiments using these fibers typically achieve spectral efficiencies of about 200 b/s/Hz, which is about 1/2–1/3 of the product of "the highest spectral efficiency in SMF (17.3 b/s/Hz)" and "the SDM count (30–40)." MM-MCF, with a SDM count exceeding 100, achieves spectral efficiencies exceeding 1,000 b/s/Hz. Since the combined bandwidth of the C-band and L-band, widely used in optical fiber communications, is approximately 10 THz, a spectral efficiency of 1,000 b/s/Hz enables a transmission capacity of 10 Pb/s.

Figure 9.21 shows an example configuration of an SDM transmission system using MM-MCF. This example illustrates the transmission of X channels of WDM signals through an N-mode M-core fiber. At the transmitter, the optical signal is multiplexed in the order of wavelength, mode and core multiplexing. At the receiver, the signal is demultiplexed in the reverse order: core, mode and wavelength demultiplexing. However, when the modes are coupled, demultiplexing using only a mode demultiplexer is not possible, and MIMO signal processing is required. In some cases, both modes and cores may be coupled, but this would result in enormous computational processing, so it is generally assumed that the cores are uncoupled.

In SDM transmission systems, the required number of the transponders is the product of "WDM count" and "SDM count." When the SDM count is increased to over 100 through MM-MCF transmission, and assuming WDM count of about 100 similar to conventional WDM systems, more than 10,000 transponders would be needed. Therefore, from the perspectives of space and power consumption, the aggregation and integration of transmitters and receivers become extremely important.

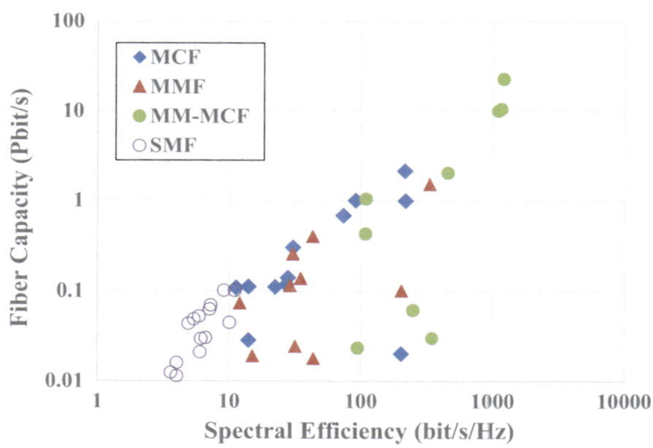

Fig. 9.20 Spectral efficiency and fiber capacity in large-capacity transmission experiments

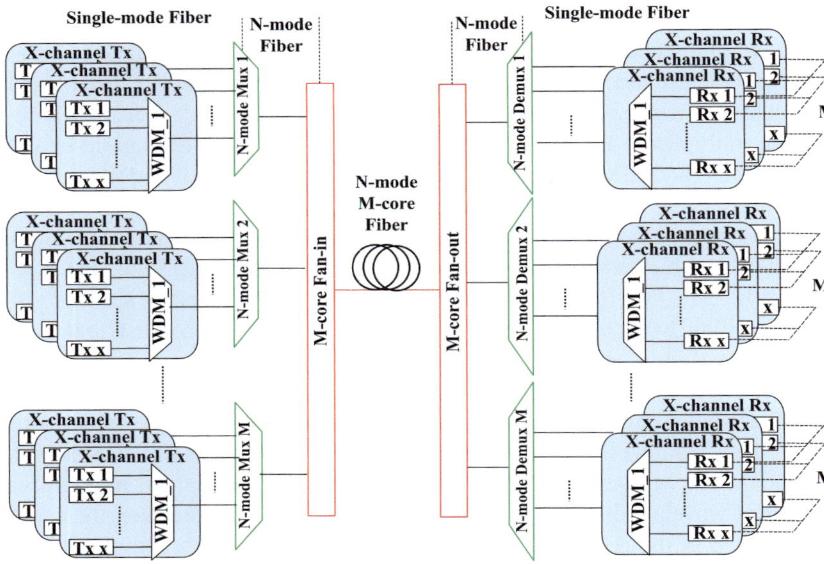

Fig. 9.21 SDM transmission system using MM-MCF

To date, SDM counts exceeding 100 have been achieved with 3-mode 36-core fiber (SDM count: 108) [10, 11, 26], 6-mode 19-core fiber (SDM count: 114) [27, 28], and 10-mode 12-core fiber (SDM count: 120) [29]. The first transmission with a fiber capacity exceeding 10 Pb/s was demonstrated with 6-mode 19-core fiber in 2017 [9] and six years later, the highest capacity was doubled to 22.9 Pb/s with 3-mode 36-core fiber [11]. The first 10 Pb/s transmission experiment is explained in the following.

Figure 9.22 shows a cross-sectional photo and refractive index profile of the 6-mode 19-core fiber used in this transmission experiment. The core diameter of this optical fiber is 18.4 μm, the core spacing is approximately 51 μm, and the cladding diameter is 267 μm. By implementing a trench structure, the core-to-core crosstalk in the 19-core fiber with a length of 11.3 km was suppressed to below − 50 dB. Additionally, the refractive index profile of the core section was designed with a graded-index type, allowing for the propagation of 6 modes: LP_{01}, LP_{11a}, LP_{11b}, LP_{02}, LP_{21a}, and LP_{21b}, and the DMD was suppressed to below 0.48 ns/km.

In this experiment, polarization-multiplexed 64QAM signals or 16QAM signals with 12 Gbaud were multiplexed at a 12.5 GHz spacing in both the C-band and L-band, and the transmission distance was 11.3 km. Among the 739 channels of wavelength-multiplexed signals, 67 channels around the shortest and longest wavelengths did not provide a sufficient signal-to-noise ratio due to insufficient gain from the optical amplifiers used in the transmitter and receiver. As a result, 16QAM signals, rather than 64QAM, were used for these channels.

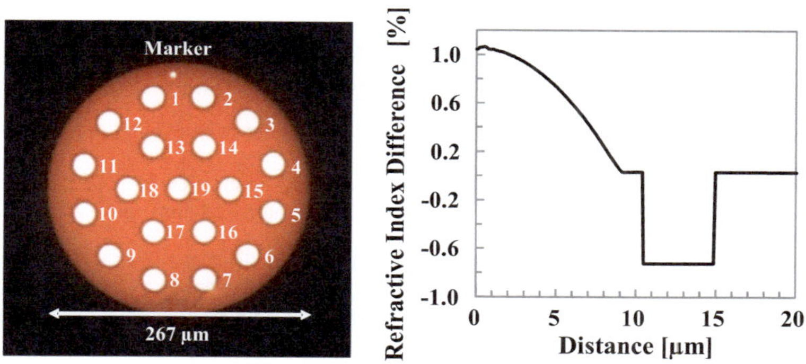

Fig. 9.22 Cross-sectional photograph and refractive index profile of 6-mode 19-core fiber used in 10 Pb/s transmission experiment

Regarding FEC codes, three types of FEC codes (with redundancy of 12.75, 20, and 25.5%) were assumed, and the optimal FEC code is selected based on the transmission performance of the signal.

By selecting the optimal modulation format and FEC code according to the transmission performance, it has been confirmed that signal performance exceeding the FEC threshold can be achieved for all 84,246 channels, corresponding to 739 channels × 19 cores × 6 modes. As a result, a fiber capacity of 10.16 Pb/s and a spectral efficiency of 1099.9 b/s/Hz have been achieved. This result updates the previous maximum transmission capacity by more than five times, indicating that optical communication transmission capacity has reached the "Ten quadrillion (10 Pb/s)" era.

Problems

9.1 Calculate the required signal power for increase the spectral efficiency from 1 to 2 or 4 b/s/Hz using Eq. 9.1.2 assuming noise power is 0.01 mW.

9.2 Calculate the required total signal power to increase the spectral efficiency from 1 to 2 or 4 b/s/Hz in SDM systems by increasing the SDM channels from 1 to 2 or 4 using Eq. 9.2.1. Assuming noise power is 0.01 mW and the total signal power is proportional to the number of SDM channels.

9.3 Explain the technologies to reduce the crosstalk between cores in multi-core fibers.

References

1. R.-J. Essiambre, G. Kramer, P. Winzer, G. Foschini, B. Goebel, Capacity limits of optical Fiber networks. IEEE J. Lightwave Technol. **28**(4), 662–701 (2010)

2. A. Sano, H. Masuda, T. Kobayashi, M. Fujiwara, K. Horikoshi, E. Yoshida, Y. Miyamoto, M. Matsui, M. Mizoguchi, H. Yamazaki, Y. Sakamaki, H. Ishii, 69.1-Tb/s (432 × 171-Gb/s) C- and extended L-band transmission over 240 km using PDM-16-QAM modulation and digital coherent detection, in *OFC/NFOEC2010, PDPB7* (2010)
3. K. Watanabe, T. Saito, K. Imamura, M. Shiino, Development of fiber bundle type fan-out for multicore fiber, in *OECC2012, 5C1-2* (2012)
4. Y. Tottori, T. Kobayashi, M. Watanabe, Low loss optical connection module for seven-core multicore fiber and seven single-mode Fibers. IEEE Photon. Technol. Lett. **24**, 1926–1928 (2012)
5. P. Mitchell, G. Brown, R. Thomson, N. Psaila, A. Kar, 57 channel (19 × 3) spatial multiplexer fabricated using direct laser inscription, in *OFC 2014, M3K.5* (2014)
6. N. Fontaine, R. Ryf, J. Bland-Hawthorn, S. Leon-Saval, Geometric requirements for photonic lanterns in space division multiplexing. Opt. Express **20**(24), 27123–27132 (2012)
7. G. Labroille, B. Denolle, P. Jian, P. Genevaux, N. Treps, J.-F. Morizur, Efficient and mode selective spatial mode multiplexer based on multi-plane light conversion. Opt. Express **22**(13), 15599–15607 (2014)
8. K. Igarashi, Y. Wakayama, D. Soma, T. Tsuritani, I. Morita, K. Park, J. Ko, B. Kim, Low loss and low-crosstalk all-fiber-based six-mode multiplexer and demultiplexer for mode-multiplexed QAM signals in C-band, in *OFC2018, Th1K.3* (2018)
9. D. Soma, Y. Wakayama, S. Beppu, S. Sumita, T. Tsuritani, T. Hayashi, T. Nagashima, M. Suzuki, H. Takahashi, K. Igarashi, I. Morita, M. Suzuki, 10.16 peta-bit/s dense SDM/WDM transmission over low-DMD 6-mode 19-core fibre across C + L band, in *ECOC2017, Th.PDP.A.1* (2017)
10. G. Rademacher, B. Puttnam, R. Luís, J. Sakaguchi, W. Klaus, T. Eriksson, Y. Awaji, T. Hayashi, T. Nagashima, T. Nakanishi, T. Taru, T. Takahata, T. Kobayashi, H. Furukawa, N. Wada, Highly spectral efficient C + L-band transmission over a 38-core-3-mode Fiber. IEEE J. Lightwave Technol. **39**, 1048–1055 (2021)
11. B. Puttnam, M. van den Hout, G. Di Sciullo, R. Luis, G. Rademacher, J. Sakaguchi, C. Antonelli, C. Okonkwo, H. Furukawa, 22.9 Pb/s data-rate by extreme space-wavelength multiplexing, in *ECOC2023, Th.C.2.1* (2023)
12. T. Mizuno, K. Shibahara, F. Ye, Y. Sasaki, Y. Amma, K. Takenaga, Y. Jung, K. Pulverer, H. Ono, Y. Abe, M. Yamada, K. Saitoh, S. Matsuo, K. Aikawa, M. Bohn, D. Richardson, Y. Miyamoto, T. Morioka, Long-haul dense space-division multiplexed transmission over low-crosstalk heterogeneous 32-core transmission line using a partial recirculating loop system. IEEE J. Lightwave Technol. **35**(3), 488–498 (2017)
13. K. Igarashi, T. Tsuritani, I. Morita, Y. Tsuchida, K. Maeda, M. Tadakuma, T. Saito, K. Watanabe, K. Imamura, R. Sugizaki, M. Suzuki, 1.03-exabit/s km super-Nyquist-WDM transmission over 7,326-km seven-core fiber, in *ECOC2013, PD3.E.3* (2013)
14. O. Sinkin, A. Turukhin, Y. Sun, H. Batshon, M. Mazurczyk, C. Davidson, J. Cai, W. Patterson, M. Bolshtyansky, D. Foursa, A. Pilipetskii, SDM for power-efficient undersea transmission. IEEE J. Lightwave Technol. **36**(2), 361–371 (2018)
15. A. Turukhin, H. Batshon, M. Mazurczyk, Y. Sun, C. Davidson, J.-X. Cai, O.V. Sinkin, W. Patterson, G. Wolter, M. Bolshtyansky, D. Foursa, A. Pilipetskii, Demonstration of 0.52 Pb/s potential transmission capacity over 8,830 km using multicore fiber, in *ECOC2016, Tu.1.D.3* (2018)
16. Y. Kawaguchi, T. Tsuritani, Ultra-long-haul multicore fiber transmission over 5,000 km using cladding pumped seven-core EDFA, in *OECC2017, 3-1K-3* (2017)
17. Y. Tsuchida, K. Maeda, K. Watanabe, K. Takeshima, T. Sasa, T. Saito, S. Takasaka, Y. Kawaguchi, T. Tsuritani, R. Sugizaki, Cladding pumped seven-core EDFA using an absorption-enhanced erbium doped fibre, *ECOC2016, M2A2* (2016)
18. T. Matsui, T. Kobayashi, H. Kawahara, E.L.T. de Gabory, T. Nagashima, T. Nakanishi, S. Saitoh, Y. Amma, K. Maeda, S. Arai, R. Nagase, Y. Abe, S. Aozasa, Y. Wakayama, H. Takeshita, T. Tsuritani, H. Ono, T. Sakamoto, I. Morita, Y. Miyamoto, K. Nakajima, 118.5 Tbit/s transmission over 316 km-long multi-core fiber with standard cladding diameter, in *OECC2017, PDP-2* (2017)

19. D. Soma, S. Beppu, Y. Wakayama, S. Sumita, H. Takahashi, N. Yoshikane, I. Morita, T. Tsuritani, M. Suzuki, Trans-Pacific class transmission over a standard cladding ultralow-loss 4-core fiber. Opt. Express **30**, 9482–9493 (2022)
20. https://cloud.google.com/blog/products/infrastructure/delivering-multi-core-fiber-technology-in-subsea-cables?hl=en
21. R. Ryf, J. Alvarado-Zacarias, S. Wittek, N. Fontaine, R. Essiambre, H. Chen, R. Amezcua Correa, H. Sakuma, T. Hayashi, T. Hasegawa, Coupled core transmission with 7-core fiber, in *OFC2019, Th4B.3* (2019)
22. G. Rademacher, R. Lu'ıs, B. Puttnam, N. Fontaine, M. Mazur, H. Chen, R. Ryf, D. Neilson, D. Dahl, J. Carpenter, P. Sillard, F. Achten, M. Bigot, J. Sakaguchi, H. Furukawa, 1.53 peta-bit/s C-band transmission in a 55-mode fiber, *ECOC2022, Th3C.3* (2022)
23. R. van Uden, C.M. Okonkwo, H. Chen, H. de Waardt, A. Koonen, Time domain multiplexed spatial division multiplexing receiver. Opt. Express **22**(10), 12668–12677 (2014)
24. D. Soma, S. Beppu, Y. Wakayama, Y. Kawaguchi, K. Igarashi, T. Tsuritani, 257-Tbit/s partial MIMO-based 10-mode C + L-band WDM transmission over 48-km FMF, in *ECOC2017, M.2.E.3* (2017)
25. S. Olsson, J. Cho, S. Chandrasekhar, X. Chen, E. Burrows, P. Winzer, Record-high 17.3-bit/s/Hz spectral efficiency transmission over 50 km using probabilistically shaped PDM 4096-QAM, in *OFC2018, Th4C.5* (2018)
26. J. Sakaguchi, W. Klaus, J. Mendinueta, B. Puttnam, R. Luís, Y. Awaji, N. Wada, T. Hayashi, T. Nakanishi, T. Watanabe, Y. Kokubun, T. Takahata, T. Kobayashi, Large spatial channel (36-core $\times$ 3 mode) heterogeneous few-mode multicore fiber. J. Lightwave Technol. **34**(1), 93–103 (2016)
27. T. Hayashi, T. Nagashima, K. Yonezawa, Y. Wakayama, D. Soma, K. Igarashi, T. Tsuritani, T. Sasaki, 6-mode 19-core fiber for weakly-coupled mode-multiplexed transmission over uncoupled cores, in *OFC2016, W1F.4* (2016)
28. T. Sakamoto, T. Matsui, K. Saitoh, S. Saitoh, K. Takenaga, T. Mizuno, Y. Abe, K. Shibahara, Y. Tobita, S. Matsuo, K. Aikawa, S. Aozasa, K. Nakajima, Y. Miyamoto, Low-loss and low-DMD 6-mode 19-core fiber with cladding diameter of less than 250 μm. J. Lightwave Technol. **35**(3), 443–449 (2017)
29. T. Sakamoto, K. Saitoh, S. Saitoh, Y. Abe, K. Takenaga, A. Urushibara, M. Wada, T. Matsui, K. Aikawa, K. Nakajima, 120 spatial channel few-mode multi-core fibre with relative core multiplicity factor exceeding 100, in *ECOC2018, We3E.5* (2018)

MIX
Papier aus verantwortungsvollen Quellen
Paper from responsible sources
FSC® C105338

If you have any concerns about our products,
you can contact us on
ProductSafety@springernature.com

In case Publisher is established outside the EU,
the EU authorized representative is:
Springer Nature Customer Service Center GmbH
Europaplatz 3, 69115 Heidelberg, Germany

Printed by Libri Plureos GmbH
in Hamburg, Germany